Agriculture in the Third World

A Spatial Analysis

Westview Advanced Economic Geographies

Land Reform
Russell King

Agriculture in the Third World: A Spatial Analysis
W. B. Morgan

Agriculture in the Third World

A Spatial Analysis

W. B. Morgan

Westview Press
Boulder, Colorado

Westview Advanced Economic Geographies

Published in 1978 in the United States of America by
Westview Press, Inc.
1898 Flatiron Court
Boulder, Colorado 80301
Frederick A. Praeger, Publisher and Editorial Director

Library of Congress Cataloging in Publication Data
Morgan, William Basil.
Agriculture in the Third World.
(Westview advanced economic geography)
Bibliography: p.
Includes index.
1. Underdeveloped areas—Agriculture.
2. Agricultural geography. I. Title. II. Series.
HD1417.M68 1978 338.1'09172'4 78-24064
ISBN 0-89158-820-5

Acknowledgments

I wish to acknowledge the encouragement and advice of Professor R. O. Buchanan, as series editor, who suggested I write this book. His effort has made it much more readable than would otherwise have been the case. I am indebted to the cartographers of the King's College Geography Department for their skill in producing maps and diagrams from my original data and sketches. My wife, Joy, typed and retyped the manuscript. To her patience I owe a great deal.

I wish to thank the following for permission to use material already published: The Oxford University Press for data from the *Oxford Economic Atlas of the World*, 4th edn., 1972 (figs. 14, 15, 16, 17, 19); Geographical Publications Ltd. for material from D. N. McMaster, *A Subsistence Crop Geography of Uganda*, 1962 (fig 22); Hodder and Stoughton for material from K. M. Barbour, *The Republic of the Sudan*, 1961 (fig. 26); Gerald Duckworth and Co. for data from A. H. Bunting (ed.) *Change in Agriculture*, 1970 (fig. 20); and Praeger Publishers Inc. for data from G. E. Schuh, *The Agricultural Development of Brazil*, 1970 (fig. 21).

My thanks also to Tony Allan, Chris Dixon, Don Funnell, Tony Lemon, Tony Plumbe and David Preston for permission to quote from unpublished papers.

Contents

Maps and Diagrams

Tables

Preface

'. . . we do not yet seem to have realised that the exchange of products between countries in one part of the world but at different stages of development is no less natural, and no less profitable for the various nations, than the exchange of products which differ because they grow in different climates' (Thünen-Hall, 1966, p. 194).

There have been few attempts to study agriculture within a spatial framework, notwithstanding the quintessential importance of land as a production factor. Land is most often treated as generalized environment although it could also be considered as social and economic space–social because even the most crowded of farming communities has much greater distance between its basic social units than exist within an urban-industrial agglomeration, and economic because distances to markets, to factor sources and to information must be overcome and frequently vary by type of market, factor and information source. Modern agricultural geography has been largely preoccupied with the development of techniques and with classification, often as ends in themselves, or with a geographical element consisting mainly of some general locational reference or regional description. Rarely has there been an attempt to identify a spatial structure associated with some particular agricultural enterprise* or practice.

* 'Enterprise' here refers to farm product for sale such as a particular crop or livestock product, e.g. wheat or beef. This use is frequent amongst agricultural economists and agricultural geographers and differs from the normal meaning of 'enterprise' in economics.

Space acts as a filter for economic activity. Accordingly we can identify certain kinds of agricultural practice as associated with particular locations in the framework of social and economic space. There is a pattern of world agriculture which suggests that economies at a low level of development are frequently associated with agricultural activity sufficiently distinctive to be worthy of special study. The theme therefore is the identification and explanation of spatial pattern in a particular sector of the spatial economy. In the Third World, usefully distinguished from the 'centrally planned economies' for correspondence of economic organization and for ease of obtaining data—United Nations' statistical tables identify the Third World as 'developing market economies' (United Nations Food and Agriculture Organization, 1973 and appendix)—this identification applies to the largest sector of the economies of most countries concerned and for many of them the most vital for development potential.

Spatial relationships can embrace the entire earth or at another extreme can be limited to the pattern of crops and of farm work in a single field or plot. An approach to agricultural geography by study at different scale levels has been attempted by Chisholm (1962) in a thesis embracing farm, village, region and world, and by Norcliffe (1969) in the study of a small region. The general problems of study at different scales are fundamental in geography (Haggett, 1965A, pp. 263–5; Haggett, 1965B; Harvey, 1968) and have been discussed elsewhere (Morgan and Munton, 1971, pp. 3–4). In examining agriculture in a framework of economic development it is essential to understand how different are the factor relationships at the different scale levels and even the nature of the data used for study and the kinds of question a research worker may ask. No great understanding will be achieved by sampling at a small unit level such as the farm and then generalizing at a higher level. The fallacy of composition, i.e. that that which is true of the part is not necessarily true of the whole, has been discussed with reference to this problem of size of unit in regional classification by Chisholm (1964).

To avoid this danger and in part because the general reasons for choice of a particular range of farm systems relate to location in the system of world trading and political relationships, the examination of Third World agriculture will begin at the world level to identify the main elements in the general pattern. They will proceed by national, regional and local core and market areas to the farm level. It is recognized that the distinctions between the different scale levels are not easy to make, nor is it claimed that other choices of level may not be made. Sizes differ in different locations and therefore one cannot characterize each scale by a G-value (Haggett, Chorley and Stoddart, 1965) except in average terms which in specific cases may have little relevance. Here region is used in a special sense to refer to the concept of 'agricultural region' as a broad organized area of agricultural production distinguished by enterprise specialization or by concentration on some particular practice. Since it is organized, it is normally sub-national and will be so regarded here, although regions of various kinds have of course been recognized at most scales. For example, there are multinational units within which common policies regarding agriculture or regarding a particular crop prevail. Most of these units are of recent origin and within most of them the attempts to develop unified market arrangements or a single system of control over agricultural production have occasionally failed or where successful have often had only limited geographical effects. In a few cases we can recognize strong geographic controls, observation of which has not been treated separately but has been incorporated into the world scale analysis (see pp. 94–5, 113–16). One may note that in the formation of multinational economic groups distance has been shown to be only one of several variables, some of which appear to have had greater influence, especially the size, cultural and structural variables (Wolf Jr. and Weinschrott, 1973).

I

Introduction

Agriculture in the Third World has frequently been denounced as barbarous, primitive and wasteful in its use of resources. It has, however, not lacked its apologists or its defenders, who have explained existing practice in terms of responses to environment learned over generations or in terms of social and economic constraints, and who have warned of the dangers of upsetting existing agricultural systems through agricultural 'improvement'. Whilst the first view tends to assume that a scientifically based agriculture of a kind associated essentially with the industrial nations of Europe or North America must be superior in any situation, the second tends to make sacred any apparent achievement of general environmental and economic equilibrium by an agricultural system and may inhibit change even where change is both possible and desirable. The agricultural improvers share with some theorists of 'modernization' schools in economics and geography a general view that the latest innovation must spread everywhere that is environmentally possible or, as Thünen described the situation in 1826: 'An ancient myth pervades our agricultural writings: that whatever the stage of social development, there is one valid farming system only—as though every system that is more simple, every enterprise that adopts extensive methods to economise on labour, were proof of the practising farmer's ignorance' (Thünen–Hall, 1966, p. 258).

Thünen's originality was not the recognition of a relationship between farming practice and distance from market or farmstead, but the recognition of a dynamic situation in agriculture in relation to market and the development of a simple apparently static model to describe it. Thünen's rings are innovation rings and his model implies not only locational differentiation of farming but a spatial pattern of change. Thünen never envisaged that in reality the zones would be circular and that innovation would always move outwards. Such appearances are properties of the model designed in this manner for the sake of simplicity. Nevertheless contiguity is clearly an important element in the dissemination of information, as is also the sharing of common information-spreading networks such as a telephone system or news media organization such as Reuters. As a result many innovations have tended in practice to spread outwards. A new practice belongs, however, to the socio-economic environment in which it was created. It must solve a particular problem and can be spread only in so far as there exist over a large area similar socio-economic environments or environments in which there are related problems for the solution of which the innovation may be adapted. British mixed farming systems would be as unsuitable in the Brazilian highlands or coastal lowlands as the fazenda systems would be in Britain. In the process of diffusion an innovation may change its form considerably, as, for example, in the development of large-scale plantation systems of production in the Third World. These in one sense may be regarded as the introduction of techniques of European origin, but in another sense are an innovation different in character from any large-scale agricultural practices ever found in Europe (see pp. 72–4). Modernization may and often does assume peculiar forms in those areas of the Third World most open to European or North American penetration.

THE WORLD AGRICULTURAL SYSTEM

Peet (1969) has summarized and developed ideas of a world-scale agricultural zonal system (Chisholm, 1962, pp. 189–90) focused on 'world metropolis' (Schlebecker, 1960) and spreading outwards, so that an agricultural frontier has been created which invades continental interiors as the outer boundary of a dynamic system whose movements are explicable in terms of changes in internal supply and demand. This world agricultural 'system' is, however, not the simple structure suggested by the notion of world metropolis, but a highly complex set of markets and of economic distribution and production systems, in which in the most developed economies almost all agricultural produce is distributed through centralized markets, whilst in the least developed economies most agricultural produce is consumed by the farmers themselves and their dependants. The world 'system' has not one but many centres of demand and input supply, each of a somewhat different character and subtending different trading areas. Growth has not only spread outwards but has increased the number of centres and the complexity of their trading relationships. In the most developed economies increased agricultural production has mainly been achieved by intensification and has frequently been accompanied by a decrease in the area under cultivation. In the least developed economies increased agricultural production was the product until 1966 of increases in the area farmed, and even since then increased production through higher yields achieved by greater inputs has been characteristic of only a few countries with especially favourable conditions for the planting of high yielding varieties of wheat, rice and, to a lesser extent, maize and sorghum. In the rice producing regions, especially, most of the production increases achieved before the introduction of the 'new seeds' were associated either with shifts in the regional distribution of the area devoted to rice or with modification of the environmental factors (Ruttan, 1968). Generally in the Third World the level of agricultural technology and the factor proportions in agriculture, despite considerable effort to improve productivity, have remained relatively constant and have tended to increase mainly in proportion to the increase in the agricultural population and in the workforce available as additional land has been brought into cultivation (McPherson, 1968). In very broad terms a real contraction has occurred at the 'centres' of the world agricultural system and expansion at its periphery. A remarkable feature of the Third World has been the mobility of its rural settlement and the speed with which transport and trade infrastructures have been extended to cope with the increased area under commercial crops. These changes are not, however, necessarily indicative of peripheral vigour in agriculture. On the contrary, peripheral expansion has often been the result of a general lack of innovation or of limitation in the effectiveness of current agricultural innovations in the economic and physical environments of the Third World. In some instances gross overcrowding, as in Bangladesh or in the Ganges Plain of India, has made areal expansion impossible, but even in these areas, apart from the introduction of high yielding crop varieties, little innovation of importance has taken place, and periods of famine have been relieved by the import of food grains from more developed countries. The Third World has thus suffered from its peripheral position, and the major agricultural innovations that have succeeded there have been for improvements in the production of industrial crops for the expanding markets of the more developed countries. The demand for these has proved much more elastic than the demand for food in the less developed economies, despite the generally agreed need for increase in both the quantity and the quality of the latter. In the more developed countries agricultural production has concentrated mainly on the needs of regional or national markets, more especially for food crops. Some countries, such as Denmark and Ireland, have specialized in satisfying certain food demands of other more developed countries. Related forms of agricultural production have appeared at the apparent periphery of the world agricultural system in Argentina, Australia and New Zealand. But the absolute geographical relationships are deceptive, for these countries provide broad environmental analogues of the temperate conditions characteristic of most of the more developed world. They have provided outlets for European emigration and have been favoured by relatively high levels of capital investment. The real periphery is not here, but almost entirely in the tropics and sub-tropics.

THE TROPICAL WORLD

Much of the Third World is tropical (Fig. 1); that is, it is frost free, has only small variations in the length of day, has seasons defined by water availability rather than by temperature,

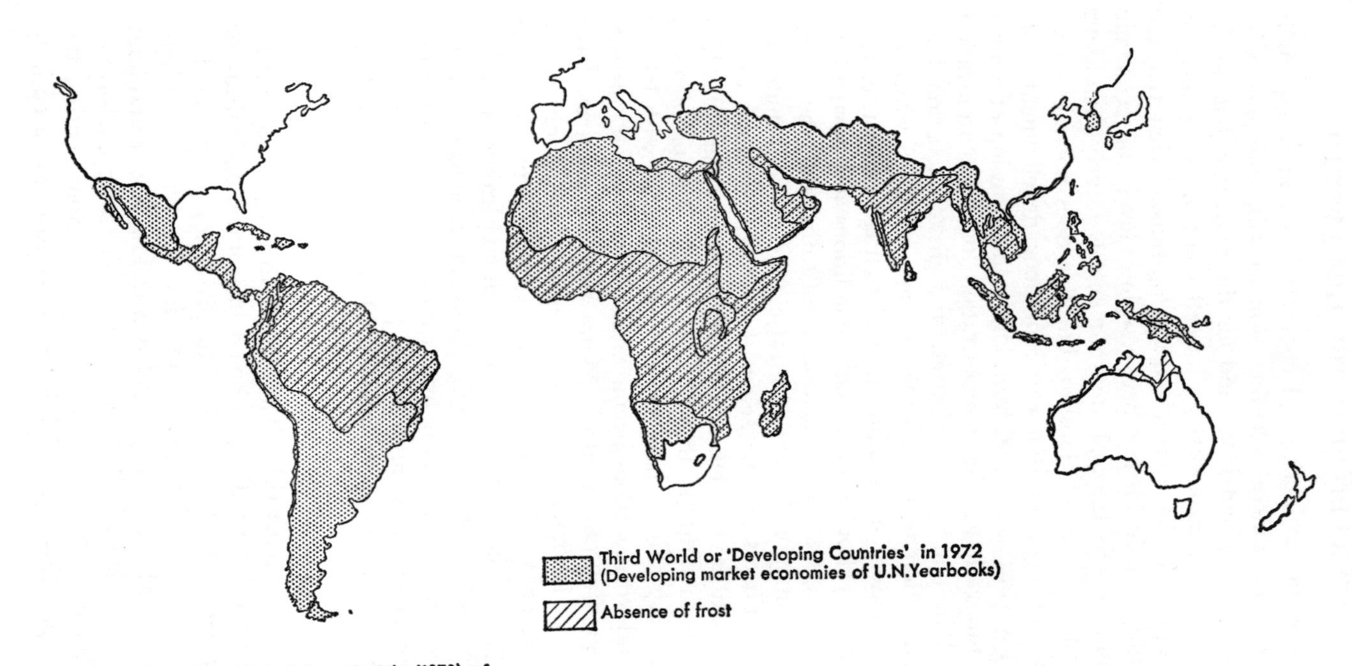

Fig. 1. The Third World and absence of frost

possesses some distinctive soil characteristics, more especially high levels of mineral accumulation in soil, combined with rapid leaching of nutrients, and for the most part has special difficulties of disease, pest and weed control. Seers and Joy (1971, pp. 78–9) have criticized the 'optimistic' view that the geographical coincidence between low levels of economic development and tropical environments is largely fortuitous. Clearly there is a relationship, but this does not mean that tropical environment is a cause of poor development. The tropical world has suffered from lag in the diffusion of appropriate innovations, the concentration of most innovation potential in temperate countries, and distant location both from innovation sources and from major markets, which has helped to reduce profitability and discourage enterprise. It is not that very hot or never cold conditions are particularly difficult for agricultural innovation, but rather that the attractions of the market for agricultural equipment and materials, such as fertilizers and pesticides, have been so much less in tropical countries that few attempts to overcome the difficulties have been thought worth while. Moreover, the problems of temperate agriculture have been more concerned with labour saving and with location-specific problems such as soil nutrients and weed and pest control. Many tropical countries have tended to be labour rich in general terms, although often with local shortage of labour arising from structural problems (see pp. 26–7), and have suffered more from constraints of capital or, less generally, of land. Again the main impetus to agricultural development has come from overseas markets, and these have fluctuated enormously in the prices offered for tropical produce, whilst temperate foodstuffs have frequently enjoyed protected or guaranteed markets. The most attractive sector of agriculture in tropical countries in production potential is subject to enormous risks. The degree to which people from overseas have been willing to invest in tropical agriculture or local farmers have been willing to plant new crops has frequently been astonishing, given the risks involved. Only high levels of reward, frequently coupled with low labour and land costs, have made such development possible (see pp. 93–106).

For Pierre Gourou there is a distinct regional entity, the humid tropics (Fig. 2), which is seen to consist almost entirely

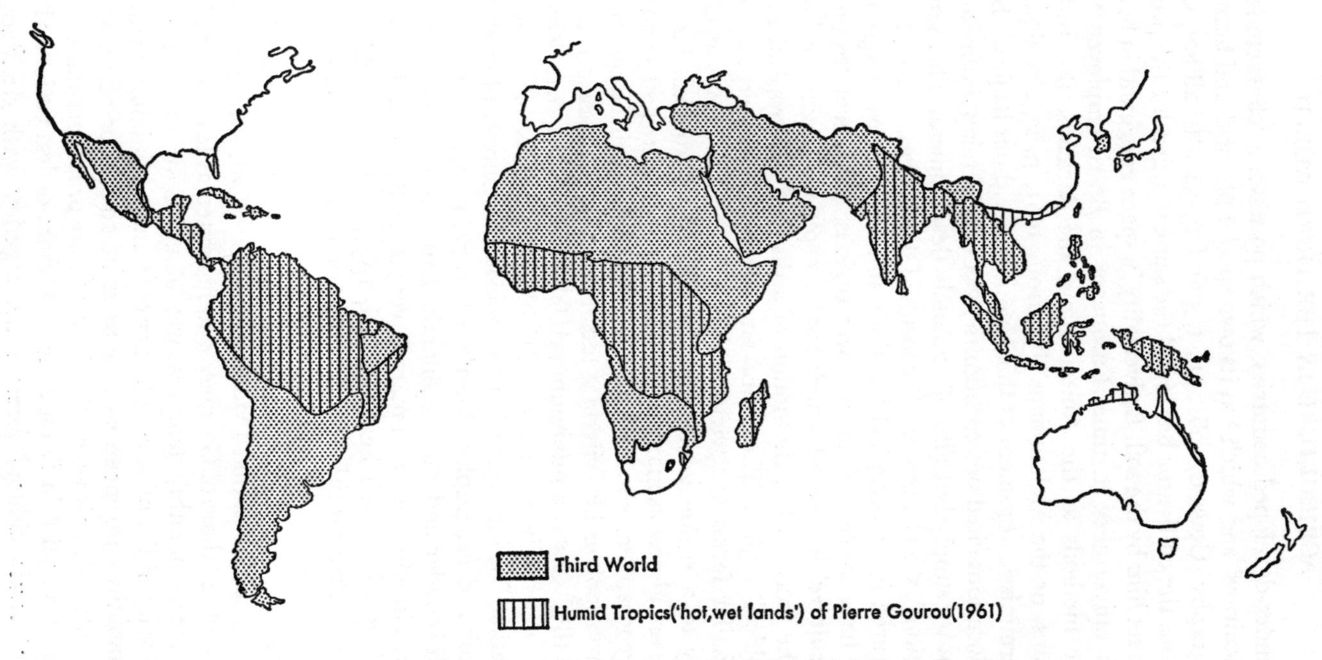

Fig. 2. The Third World and the 'hot, wet lands' of Pierre Gourou

of under-developed countries, which possesses a 'characteristic agriculture' and which has its own special physical and human geography (Gourou, 1961, pp. 1, 5 and 25–34). 'In all hot, wet regions the cultivator has found the same solution for the problems set him by the soil. Universality in space goes hand in hand with universality in time. Europeans in Brazil employed the same methods as the Amerindians, the Africans, the Indonesians, or the Melanesians' (Gourou, 1961, p. 25). A simple deterministic argument of this kind is not without its force, but the locations cited possess affinities not only of temperature and moisture supply but also of economic development. The 'same solutions', which appear to be forms of shifting cultivation, have appeared in non-tropical locations and other forms of agriculture have succeeded in hot, wet regions. The term 'tropical agriculture' has found a universal acceptance hardly justified by the facts, as has the notion of a distinctively tropical geographical region. There are few agricultural enterprises or particular forms of agricultural practice which belong exclusively to a region which may be defined as tropical, just as there are almost as many definitions of a tropical region as there are geographers who are interested in making such a definition. Gourou took as his criteria a mean monthly temperature of at least 18·3° C and a minimum rainfall of 610 mm a year, which he thought to be enough for agriculture to be possible without irrigation. Waibel, regarded by some as the founder of modern agricultural geography (Manshard, 1974, p. 2), proposed the much broader and more economic definition of the tropics as an area in which we find mature, economically valuable plants which require much heat (Waibel, 1937, p. 19), but this definition lacked precision, and many annual crops needing high temperatures can be produced in areas where the high temperatures required exist for only a short time but sufficient for their vegetative period (Manshard, 1974, p. 8). Other definitions include those of Garnier, who suggested at least 8 months with mean monthly temperatures of 20° C or more, mean relative humidity of at least 65 per cent and a minimum period of 6 months with mean vapour pressure 20 mb. or more, and of Küchler with his vegetation-based 'more or less permanently humid' tropical rain forest zone and 'more or less periodically humid' semideciduous forest zone, together with deciduous forests with relatively few epiphytes and/or savannas (Fosberg, Garnier and Küchler, 1961). There is no generally acceptable 'tropics' or 'humid tropics'. As Fosberg showed, views have tended to be so far apart that any solution to the problem of regional definition would be purely arbitrary. Convenience with regard to the definition of a hot, humid area related to the facts of agriculture suggests a preference for Gourou's more practical approach, allied to the notion, echoed by Küchler, that in the tropics 'everything which is not arid must be humid'. Accordingly for this purpose the humid tropics may be distinguished as the zone in which there is sufficient moisture supply for a rain-dependent agriculture without irrigation or resort to dry farming techniques. In practice the zone needs at least three months rainy season, the shortest growing period for the fastest maturing cereals, mainly varieties of pennisetum millet or eleusine. Limiting average rainfall totals are for the most part about 500–600 mm, although in areas of high rainfall variability higher average totals are often regarded as reasonable minima. In East Africa, for example, 760 mm has been estimated as the minimum annual requirement for cereal cultivation (Glover, Robinson and Henderson, 1954; Grigg, 1970, pp. 235–6). Rainfall variability and crop failure rate increase dramatically as rainfall totals decrease (Fig. 3). This zone may be divided into a sub-zone with varying lengths of dry season significant for agriculture and a constantly humid sub-zone which includes not only the areas with rain in every month, but even those areas with up to approximately two months 'dry' season with very high relative humidity and frequent occurrence of morning dew, often in amounts vital for plant growth. The zone also includes very large areas of irrigated and floodland agriculture.

There are several features sufficiently peculiar to agriculture in the humid tropics to be regarded as distinctive, although most of the more salient characteristics either occur elsewhere or are differences of degree rather than of kind. Distinctive features include:

1. The use of shading devices or large leaved inter-cropped shading plants such as cocoyams or bananas. For some tree crops this may take the form of a long-term succession of different shading plants providing broad leaves at different

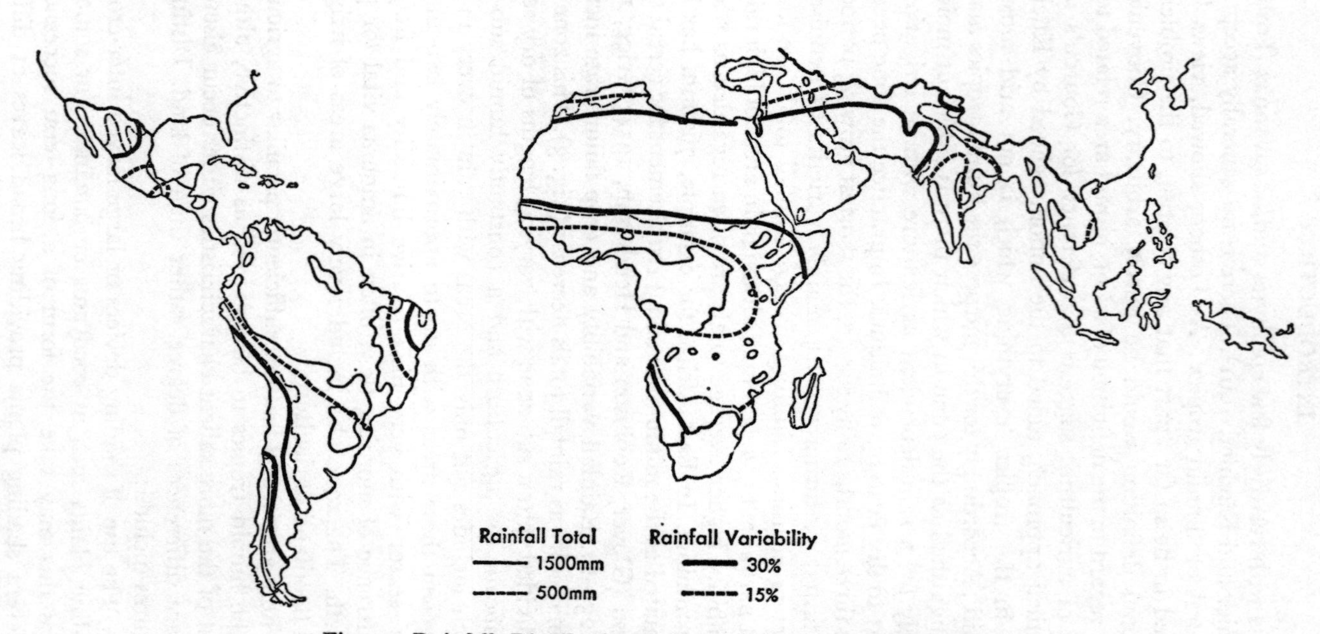

Fig. 3. Rainfall. Distribution and variability in the Third World

heights. Care has to be taken as over-shading can result in weak, thin stems (Jacob and Uexküll, 1960, pp. 84–5).

2. Rapid growth of self-sown plants including weeds and fallow plants. Some agricultural systems incur bottle necks in weeding during the growing period, others in clearance of fallow before planting (see below, p. 27).

3. High risk associated with variability in moisture supply. Highly variable rainfall occurs outside the tropics, but has a special significance in the tropics in that moisture and not temperature is the principal growth constraint. Risk avoidance techniques include production diversification and extension of the planting and harvesting periods.

4. High incidence of pests and diseases, especially diseases carried by insects. The lack of a winter break frequently exacerbates the problems. Some pests are associated with a variety of host plants, enabling them to establish themselves in a succession of plant environments in areas of marked seasonal rhythm. Control can be achieved by careful attention to fallow management and the burning of crop refuse.

5. Double cropping is rare in the temperate world, occurring there most commonly in market gardening with a few quick-growing vegetables. In the tropical world it is of common occurrence where the rainy season is nine months or more and sometimes also occurs on irrigated land. Double cropping involves planting a second crop immediately after harvesting the first and in the same location. It is to be distinguished from succession cropping where in the same field a number of crops may be planted and harvested in sequence. In some equatorial areas the constant heat and moisture supply make 'multi-cropping' possible, that is a succession of plantings, weedings and harvestings, whose seasonal rhythm is more the product of the life cycle of the plants themselves than of any rhythm of environment.

6. A marked tendency to lodging amongst cereal crops liable in hot, humid conditions to exhibit luxurious rapid growth. Optimum nitrogen dressings can in consequence be larger where temperatures are lower, resulting in a tendency for the maximum yields of crops of wide extent in the tropics, such as rice and maize, to occur in temperate regions (Jacob and Uexküll, 1960, p. 83) (Table 1). Hence the importance of

TABLE I
YIELDS OF RICE AND MAIZE (1971–3 AVERAGES) IN KG/HA

	Rice	*Maize*
North America	5117	5769
Western Europe	4823	4110
Third World	1861	1285
Third World countries in:		
Near East	3742	2364
Far East	1859	1079
Latin America	1801	1443
Africa	1387	986
World	2312	2781

Source: FAO, Production Yearbook for 1973, Rome.

attempts to develop varieties resistant to lodging in order to obtain good results from heavy nitrogen dressings in tropical conditions.

There are other features often regarded as important in tropical agriculture, which are not uncommon outside the tropics. Some of these relate to climate, such as the construction of mounds and ridges to cope with the drainage problems required by extremely heavy downpours of rain which can cause waterlogging even in areas of short rainy season with low rainfall totals. There are also the development of floodland cultivation in short rainy season areas to extend the cultivation period, the concentration on root crop staples and tolerant, mainly soft, maizes in humid environments and the development of elaborate storage techniques in areas with only short growing seasons. Others relate to soils and include the widespread occurrence of long fallows (shifting agriculture), which are a feature not only of poor soil environments but also of low input extensive systems of production. Long fallows were of not uncommon occurrence in the past in temperate countries, although today they have become rare as land values have risen. Shallow tillage occurs wherever soils are very light, unstable or liable to pan formation. Such tillage is not uncommon in temperate environments, but is widespread in the tropical world on ferruginous and sandy soiled areas with alternating wet and dry seasons liable to heavy downpours, soil erosion, leaching and the upward movement and accumulation of free sesquioxides in the soil profile. Macarthur (1971, pp. 9–10) has stressed the importance of wind in tropical environments, including the most damaging wind storms in the world, intensely drying winds, such as the West African harmattan, and a liability to wind-induced erosion in the dry season. In some locations an important role is played by shelter belts or by the avoidance of crops sensitive to wind damage, such as bananas, or by a preference for crops tolerant for a time of very dry conditions. Separation of livestock from crop raising is widespread in the tropics (contrasting with the common occurrence of mixed livestock and crop enterprises on farms in the temperate countries), and is often regarded as a distinctive feature, although it is also characteristic of Mediterranean and semi-arid lands. Partly the problem is economic and concerns limited output per unit of labour expended, with the consequent avoidance of the feed crop production characteristic of many mixed farms. Partly it is a problem of seasonality, feeding stuffs and disease. The more humid environments with long grazing seasons tend to have the fastest growing and coarsest grasses (especially perennials such as the tall *Andropogon* and *Hyparrhenia* spp.) and the highest incidence of disease. The areas with long dry seasons have better quality grazing and less disease, but require migration or irrigation to maintain water and food supplies. Some integration of livestock keeping and cropping occurs in the common rearing of small livestock in or about the farmhouse, in the use of draught animals for some cereal cultivation, in the development of a few systems which rotate pasture and crops and in the frequent grazing of nomadic livestock on dry season crop refuse.

THE NON-TROPICAL THIRD WORLD

The non-tropical, or rather non-humid-tropical, area of the Third World is mostly arid and of less agricultural importance, comprising areas mainly of pastoral nomadism, floodplain and oasis cultivation, irrigation and dry farming. Despite its smaller importance it occupies nearly 40 per cent of the area of the Third World. Essentially it has too short and too irregular a rainy season for satisfactory cultivation by normal methods. A distinction can be made between a semi-arid area with rainfall every year and with normal rainfall total limits of about 200–500 mm, in which dry farming techniques can achieve a result, and an arid area, lacking rainfall altogether in some years, where no cultivation is possible except in floodland or by irrigation, and where even nomadism is difficult to maintain away from river valleys because of the complete failure of the rains at irregular intervals. The arid areas include soil-less regions without vegetation or surface water, extensive areas with extremely light sandy soils, heavy clays subject to cracking and areas of alluvium, sometimes with permanent streams. Some arid areas are in geological basins suitable for tapping artesian water supplies or pumping water from aquifers by means of bore holes. Vegetation is generally not only sparse but includes only a limited range of species, and agriculture, likewise, except on the most favoured irrigated lands, is limited in its range of crop choice. The largest concentration of arid lands is in the Near East (approximately two-thirds of the area) and Africa, where the Sahara and Kalahari, together with smaller deserts such as the Danakil, occupy over 40 per cent of the land. Latin America and Southeast Asia have approximately a quarter of their areas in arid and semi-arid lands, including the desert lands of Mexico, the *caatinga* scrublands of northeastern Brazil, the Thar desert of India and the extensive arid lands of Pakistan. The zone includes some of the most intensive and extensive systems of agriculture and some of the most primitive and advanced farming methods in the world.

Irrigated farming is the most productive and perhaps the most distinctive form of farm practice in the arid lands, although irrigation is also widespread in wetter areas as a means of intensifying production. Irrigation makes possible multiple cropping, high yields by adequate water control, the rapid introduction of new water-demanding crop plants dependent for good results on fertilizer application, and the reduction of moisture supply risk normally characteristic of farmlands dependent on rain. It is often associated with improved techniques in tillage, because the floodplains which make up the greater part of the irrigated lands are particularly suitable for ploughing. Costs are high and most irrigation systems depend on the production mainly of commercial crops, usually of a very high value giving a high return per hectare. In less developed countries irrigation is therefore often associated with export cropping or with import substitutes, as, for example, sugar cane. Subsistence forms of agriculture dependent on surface or underground water usually involve the use of floodland with only limited attempts at management or the simplest of irrigation methods, such as shallow wells, rarely more than 10 metres deep, or the somewhat more elaborate *qanats* or underground galleries of the Middle East. The association of irrigation with the production of high yielding varieties of wheat and rice in the so-called 'Green Revolution' (see pp. 110–13) has had the peculiar effect of concentrating high productivity in lands otherwise regarded as environmentally limited, and of promoting further the movement of population to riverain areas which modern developments in hydraulic engineering have so spectacularly encouraged. The ground water supplies of the arid zone are extremely limited (Grigg, 1970, p. 188) but nevertheless they are especially important in certain countries, where they provide more than half the water requirement, as in north-west Africa and Iran. Elsewhere seepage from rivers, as in the Nile Valley and the Punjab, is tapped by wells and thus extends the irrigated area. Generally the wetter lands are the richest in ground water resources and these become especially important for dry season drinking water supplies for people and livestock and may make possible the extension of agriculture into areas with adequate rainfall for cultivation but lacking domestic water supply, as in the Ferlo of Senegal.

Floodland cultivation with little or no use of controlling dykes, earth banks or irrigation channels is a primitive use of surface water resources. It is widespread, but supports only small numbers of people, usually maintaining only high risk systems of low productivity. Characteristically it occurs mainly in the upper basins of major rivers, in small areas wherever the range of flooding is limited and the risk of very deep, damaging floods is known to be slight. It is especially characteristic of the more tropical semi-arid and savanna lands of Africa, where it often provides a supplementary production to other forms of agriculture. But it is widespread also in Southeast Asia, especially in the lower Ganges Valley and the Ganges–Brahmaputa delta and occurs in economically very developed countries where traditional practices still survive, as amongst the

Hopi Indians of Arizona (Davis, 1948, p. 410). Oasis cultivation is a special limited case of the use of surface waters occasionally supplemented by tapping ground water resources. Except for the lowland oases of the North African coastal plain, which can even suffer from excess water in the winter rains (Dumont, 1957, p. 164), oasis cultivation takes place in relatively isolated locations. These may be linked by modern motorable roads or ancient trading routes for which the oases have provided vital foci of water and food supply. In consequence whilst oasis cultivation usually concentrates on basic foodstuffs, despite its limitations and the need for intensity of production, cropping often has a commercial character and may find adequate markets. Date palm plantations are the most characteristic feature, especially in North Africa and the Middle East, and fodder crops such as barley are grown. The decline of a desert crossing traffic dependent on local resources has throughout the world brought economic decline to the oases, which today are often more important as sources of water for nomadic livestock and occasionally as the focal points of nomadic pastoral systems. Some have gained a new importance with the exploitation of minerals, more especially of oil.

Semi-arid cultivation or dry farming is the technique of producing mainly cereals in areas in which in most years the length of the rainy season and the supply of moisture are too short for normal cultivation methods. It is best developed in the cooler arid lands, more especially those with winter rains, where the water loss to evaporation is less. Land is fallowed for over a year before cultivation and frequently tilled to reduce or eliminate plant cover and prevent transpiration. Labour input is considerable and the risks of soil erosion are high. The aim is to conserve some at least of a season's moisture in the soil to supplement the moisture supply of the growing period. With high weather variability and the risks of dessication by very dry winds yields are low and the danger of disaster is considerable. Intensification is normally not worth the risk and the possibilities of innovation and improvement are poor. Indeed there is a tendency to retreat to the better watered lands.

Most of the arid land area which is subject to management is used for short periods of grazing usually on open range. Tussocky grasses, shrubs, such as the North African *betoum* or

Atlantic turpentine tree (*Pistacia atlantica*), and various small plants provide limited sustenance for migratory herds, whose range is limited not only by the availability of grazing but by the occurrence of drinking water and supplementary feeding from better quality pastures on annually flooded land. Such grazing has decreased with the decline of desert trading, the advance in some areas of arable farming and the attractions of better living elsewhere or permanent damage to some pastures from over-grazing. In the semi-arid areas the regular north to south movement of the rains and associated young grass pastures has supported a regular seasonal rhythm of grazing, which still survives, notably on the southern margins of the Sahara and in the Middle East.

A very small area of the Third World is neither arid nor humid tropical, but consists of lands with alternating hot-dry and cold-humid seasons or Mediterranean climate, the cool or cold uplands within the tropics having tropical diurnal and seasonal rhythms of temperature and light but with a great range of moisture supply, a few very small areas of cool temperate climate, mainly in South America, and finally the humid sub-tropical lands of north-east Mexico and South Korea. The largest extent of Third World Mediterranean lands is in North Africa and the Middle East, including the Turkish coastlands. Elsewhere the Mediterranean environments are 'developed', with the exception of the Mediterranean region of Chile. Third World mountain and plateau lands are extensive in the Middle East, in the high plateaux of Central America and the South American Andes and in such isolated plateaux lands as Ethiopia. Although there are some interesting links between agricultural practices in the various Mediterranean lands and between practices in the various plateaux, one is more impressed by the variety of both enterprise and practice even in very similar environments and at apparently similar levels of economic development. Most of these areas share with the tropics a lack of integration between livestock rearing and crop raising, which in the Mediterranean lands has been ascribed to the lack of a suitable legume which can be inserted into a rotation and can provide feed for livestock and a replenishment for soil fertility comparable to clover in temperate areas (Grigg, 1974, p. 125).

AGRICULTURE IN LESS DEVELOPED ECONOMIES

If it is difficult to distinguish a distinctively tropical agriculture, it may be thought equally difficult to isolate a less developed agriculture. It is contended, however, that this latter task is much less formidable and that the regionalization of agricultural characteristics associated with certain kinds of economy is clearer. This contention is made despite the existence of a general consensus identifying a tropical agriculture whilst few are interested in identifying specifically less developed forms of farming. Despite the undoubted importance of physical and biological factors in agricultural production and the frequent observation of spatial correlation between certain crops and certain sets of physical conditions, the physical and biological environments can only define the areal extent of agricultural systems indirectly through the dominance in a particular system of a feature strongly limited by environmental constraints. The biophysical environmental characteristics are not essential features of agricultural systems as such, but a framework within which the systems operate, a framework of both resources and constraints. Agriculture has other frameworks or environments, such as the society or the economy within which it operates, but essentially farms are businesses with economic objectives. Farming must satisfy the needs of subsistence or the demands of the market, and this is true even where farmers cannot know optimum solutions to their problems of enterprise choice or are satisfied with farming 'as a way of life'.

A less developed country is not a member of a discrete group but a nation which is low on various scales of comparison. In Fig. 4, for example, Third World and non-Third World countries are compared with regard to gross national product. The less developed economies are affected by a large variety of common socio-economic problems and deficiencies of indirect but important significance for agriculture. These include low levels of education, limiting agricultural training, research and development of an agricultural cadre (see opposite, Fig. 4); an education transition in which large numbers of children are no longer available for farm-work, resulting in an intensification of labour bottle-neck problems and a labour shortage

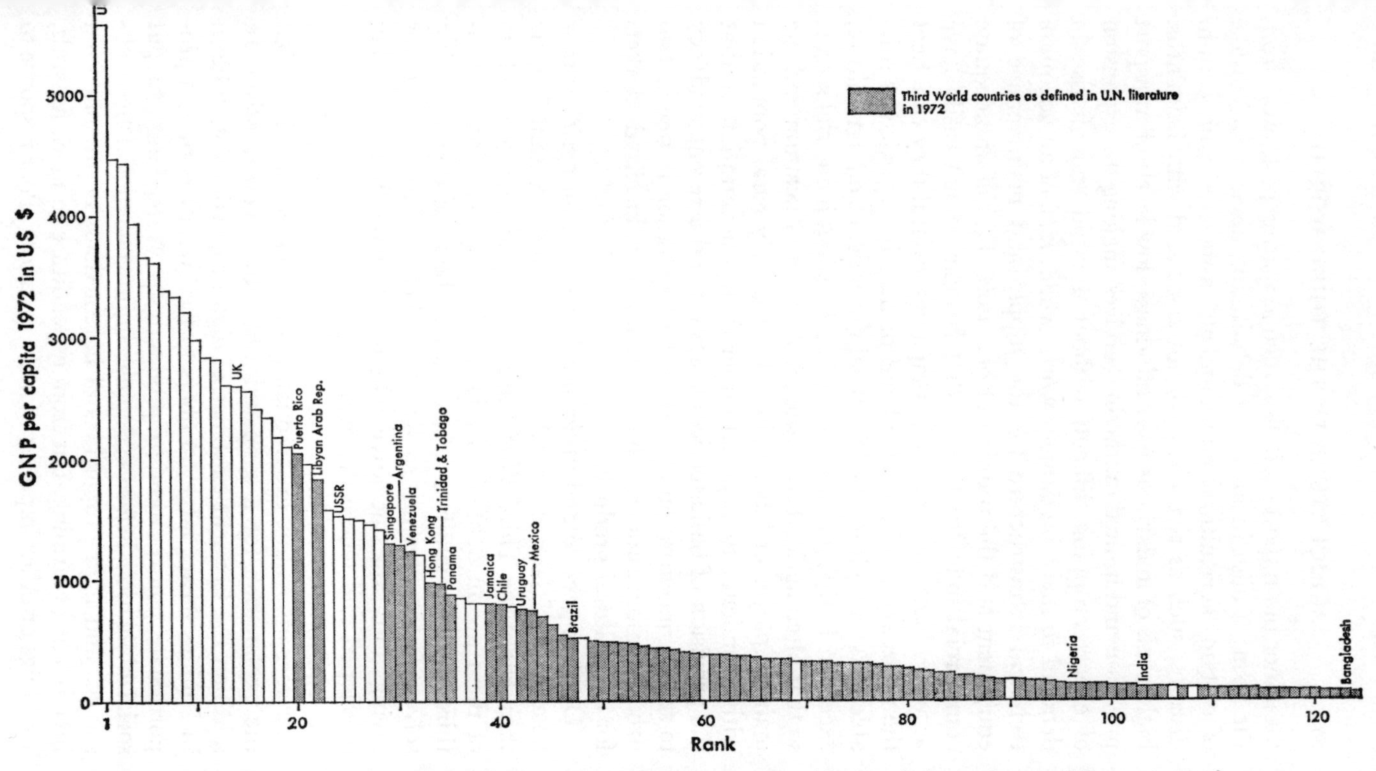

Fig. 4. Rank graph of gross national product per capita of countries with populations of one million or more in 1972

for minor but important duties, such as scaring pests away from the farm; a high incidence of debilitating diseases and dislike of certain 'agricultural environments' associated with certain diseases, such as riverain lowlands associated with helminths, high levels of malaria or river blindness; low levels of transport provision and limited marketing services, making the expansion of export cropping difficult without a rapid enough rise in demand to make investment worth while; lack of an adequate technical infrastructure for the supply and maintenance of equipment and the training of operators; lack of an adequate commercial infrastructure so that farmers cannot easily have access to banks or insurance companies, even if they can meet the necessary conditions for loans or may be regarded as suitable risks; a high level of financial dependence on agriculture, except in Latin America (Fig. 5), and a small industrial sector, so that often agriculture is the main object of taxation and the main employer of labour (Fig. 6); a mainly rural population with a tendency to dispersal rather than concentration, so that the provision of basic services is costly; and a recent tendency to super-concentration of part of the population in one or two urban agglomerations, which create highly localized markets for agricultural produce.

Of more general and profound significance for agriculture is poverty itself, lack of capital and low incomes. Peasant farmers throughout the Third World are mostly poor. Large numbers of them are in debt—in India as high a proportion as two-thirds of farm families—often without hope of ever attaining solvency. Incomes are so low that there is little margin for saving and little incentive where commercialism is weakly developed and where the availability of consumer goods and the equipment for improvement is poor. Risk taking and therefore innovation are frequently discouraged, which is often the explanation for a supposed agricultural conservatism. In addition the frequent existence of high proportions of children in families means large numbers to feed with less than a proportionate contribution to productivity. A tendency to put social obligation before economic improvement has been observed, and high proportions of indebtedness occur, not from investment in farming, but from expenditure to mark funerals, weddings or other important family occasions. Poor response to

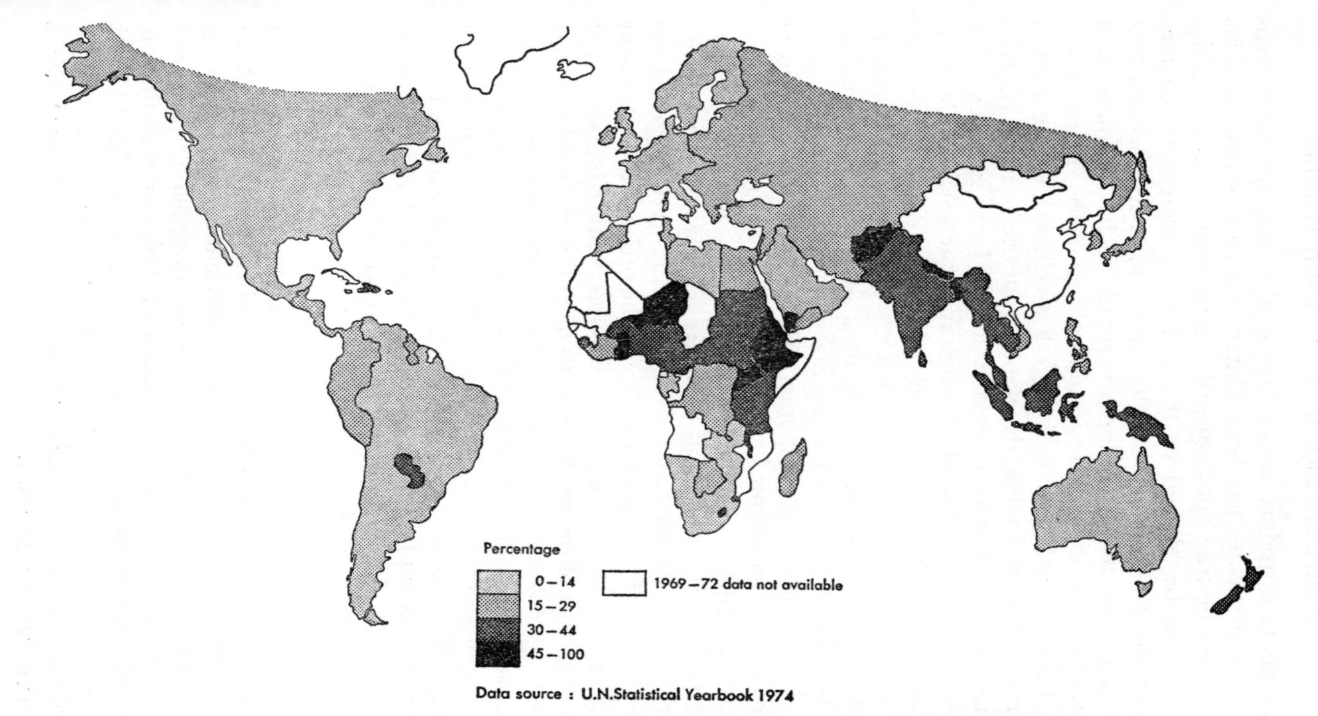

Fig. 5. Percentage contribution of agriculture to National Gross Domestic Product in 1969–72

attempts to introduce new tillage equipment, new seed, fertilizers or pesticides has been blamed on lack of capital. This may often be true but frequently uncertainty or lack of initiative should be blamed or the realization that a satisfactory return can be obtained from low fixed input systems and that the gain from new techniques is small. So far the number of possibilities for investment in agriculture in the Third World has been limited. Schatz (1965) has referred to the capital shortage illusion in less developed countries, citing Nigeria, although his examples referred solely to loans in fields other than agriculture and trade. Generally farmers have little or no collateral to raise loans, are regarded as considerable risks by money lenders, are asked for high rates of interest and tend to borrow for investment, if at all, on a short term basis, mainly for working capital to pay for variable inputs such as seed, fertilizers, sprays and temporary labour. Even for this kind of borrowing governments have frequently been driven to create special agricultural banks to provide short-term loans at low rates of interest. Credit and thrift are not always lacking, however. Pannikar (1961) and Hoselitz (1964) have suggested for India savings for farming families of the order of 8–12 per cent of income (Grigg, 1970, pp. 75–6), and Schultz has suggested that the value of the small amounts of equipment used in Third World agriculture is underestimated. If equipment is low in value so are labour and land, so that with low incomes the cost of equipment represents proportionately a considerable level of savings. Thus factor proportions in the Punjab and the United States, for example, can be shown to be similar (Schultz, 1964, pp. 99–101). There is evidence, however, of a very variable spatial situation with regard to factor proportions, so that in some areas land costs in the form of rents are very high and in others are extremely small. Poverty also leads to limitations of choice of technique, arising from lack of equipment, poverty of plant stock and inability to control nutrient supply. This is in part a reason for the similarity of systems, apparently isolated from one another, because the range of choice open to them is so small. One of the most startling features in the Third World is the universality of shifting cultivation, elevated by the typologists to the status of a system, but which is essentially a practice (see pp. 227–31).

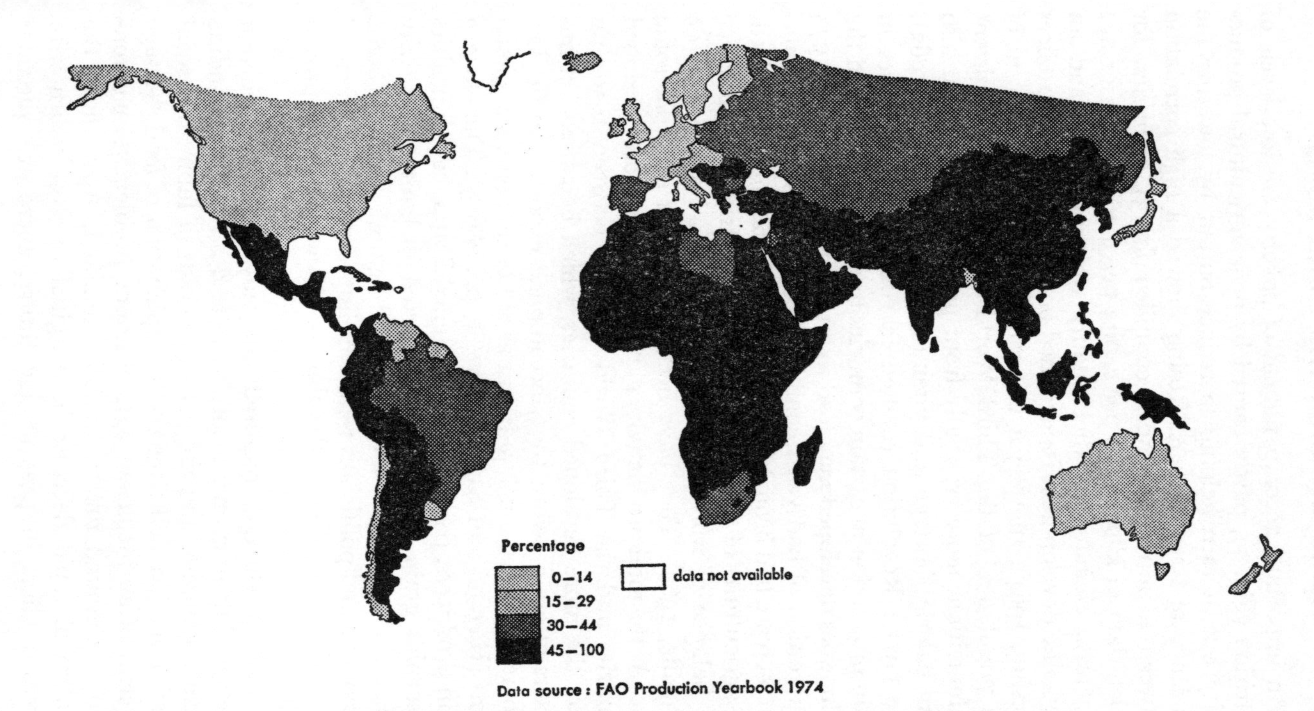

Fig. 6. Percentage of economically active population in agriculture in 1970

An important accompaniment of poverty are low levels of nutrition (Fig. 7), partly caused by poor agricultural productivity but in turn helping to reduce output by lowering the efficiency of farm workers. Rosing showed a high correlation between a nutritional index compiled for 31 countries, for which data in kilocalories per capita per day adjusted for age structure, climate and diet quality were available, and an economic development index based on fourteen specific indices (Rosing, 1964). In 1963 it was estimated that 60 per cent of the population of the Third World suffered from persistent malnutrition, more especially from protein shortage, but also from calorie shortage and vitamin deficiencies (FAO, 1963). The Green Revolution resulted in considerable increases in grain production in certain countries in the mid-1960s, and the techniques developed may lead to further progress, but by 1975 a succession of bad years combined with accelerating population increase had brought back the threat of famine. If required, the agricultures of the more developed countries could produce the balance of world food requirements for some years (de Boinville, 1968). They have been doing so in effect for some time. As the total productivity gap between the more developed countries and the Third World widened between 1950 and 1970, the more developed countries, which had been net importers of grain before 1939, became net exporters to the low income countries, much of it sent as aid or sold on concessional terms (Hayami and Ruttan, 1971, 1, pp. 286–9). The western grain producers, particularly the United States, had difficulties in surplus grain disposal and preferred to limit productivity rather than subsidize a world hedge against shortage. Some economists and politicians were over-impressed by the potential implied by the Green Revolution. Although an improved agricultural technology could increase and has already raised the food production required not only to combat current shortages but to supply a surplus to feed a rapidly expanding urban population and the potential needs of future industrial populations (Nicholls, 1963), the explanation of food shortage in terms of an inefficient agriculture or problems of environment is unsound unless carefully qualified. Apart from the problem of rapid decrease in marginal rates of return with increased effort in peasant agriculture, there are questions

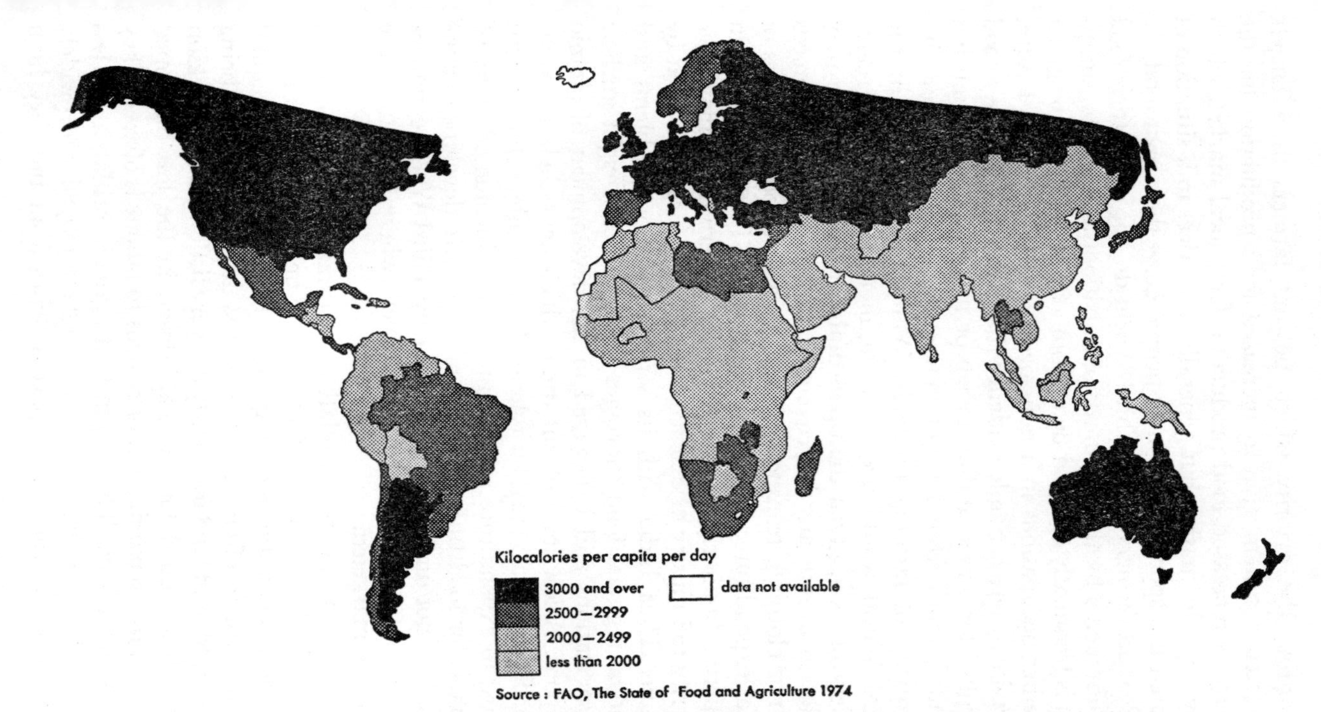

Fig. 7. Dietary energy supply in daily kilocalories per capita in 1969–71 (average)

regarding the priority which peasant farmers in a largely subsistence system give to increased food production for the family or increased food production for a local market, which may offer fluctuating but generally low prices in a situation of limited transport, where most people grow their own food.

Peasant farmers of the Third World do not judge their food requirements by the standards of nutritionists. Food consumption is frequently affected by custom and priority may be given to other occupations in a non-specialized society or to leisure. Certainly there is ample evidence that failure to provide a food 'surplus' for an expanding urban or industrial population has resulted more from lack of market incentive than from inadequacy of farming methods. Whilst many people in the Third World would prefer to eat more food and a greater variety of food, given the opportunity, nevertheless they may prefer to devote any improvement in income to other needs such as clothing, transport, saving towards important items of social expenditure, investment in business, in education or in property. Similarly at another scale governments may give priority to industry or defence, even in countries threatened by famine. Thus India with its serious nutrition problems and periodic regional food shortages today possesses eight nuclear power stations and is engaged in the development of atomic weapons. A low level of nutrition, which nevertheless is sufficient for survival, may be tolerated where other choices with regard to expenditure may be made. Low income elasticity of demand for foodstuffs exists not only in more developed countries. It is not uncharacteristic of many Third World countries despite nutrition problems and even where foodstuffs are regularly marketed.

In trying to assess the relative importance of the factors of production in Third World agriculture one finds that in many farming systems the most expensive factor is labour. Capital expenditure is often very low unless innovations are being introduced, and land frequently has very little value. As Jackson (1972) has argued for tropical Africa, in the less developed countries the maximization of returns to labour is often the key development issue. The notion that agricultural labour in less developed economies is cheap, underemployed and widely available is misleading. In part it arises from the very high proportions of labour in most less developed countries to be found in farming (Fig. 6), averaging 65 per cent of the total labour force in 1970 and declining more slowly than in the rest of the world (Grigg, 1975). Labour is often a seriously limiting factor to production at certain critical times of the year and peasant agricultural systems are normally very sensitive to marginal decline in productivity with increasing labour inputs (Clark, C. and Haswell, M. R., 1964, pp. 84–94). Leisure is often desired where labour is tedious and where certain tasks, such as forest clearance, are extremely demanding. Average farm work loads per man as low as 4 hours per day are common (Clark, C. and Haswell, M. R., 1964, pp. 117–18) despite the small output per man-hour. Most peasant farmers are non-specialized and spend in addition long hours at many other tasks. Areas with low population density tend to suffer from serious labour shortages, hampering schemes for economic development. Labour shortage has been one of the most frequent complaints of developers in tropical Africa and Latin America and has encouraged extravagant use of other inputs, notably, in Latin American latifundia systems. In labour-abundant situations labour is applied to land more intensively, especially where pressure on land resources occurs. An abundance of surplus labour is not, however, always cheaply available in such a situation and may not always be removed from agriculture without a reduction in farm output. On the contrary the labour situation has frequently resulted in the creation of systems of production with labour-demanding bottlenecks at certain seasons of the year which are very adversely affected by the loss of labour to industry or to service occupations in the towns. Only if redundancy occurs during critical periods of labour demand may the farm work force be decreased without reducing agricultural output (Jorgenson, 1966).

The notion that agricultural land in the Third World is of little value needs some qualification. It applies mainly to those less developed countries where most of the agricultural land cannot be bought and sold, but is held on a use-right or usufructuary basis. Even where land is held by usufruct, however, differences in appreciation of its capacity to produce and of its accessibility may lead to differences in land use or even of choice of occupant amongst members of the holding group.

Once the possibility of land sale is admitted, bidding may occur and prices may rise wherever commercial advantage develops. As marketing systems develop, the value of land accessible to market must rise. Rapid increases in land value have become a feature not only of a commercialized agriculture as such, but also of the competing demands of 'developers' near growing cities. Land which can be sold can be used as security for loans. Where the usufruct prevents this the risks of agricultural loans become extremely high and such loans are either not available or can be obtained only at excessive rates of interest.

Access to local or more distant, including overseas, markets is an essential feature in the development of commercial agriculture. Frequently the earliest development has been access to overseas markets through local buying stations or through pick-up points established by urban based dealers or export agents. A system for the purchase and marketing of produce involving the development of road and often also of rail transport or sometimes river traffic has to be established. Exceptionally, air transport has been used for the export of luxury perishables. Once a system has been established it can expand rapidly providing demand increases and high returns can pay the cost of the infrastructure and attract new producers. It can grow by the development of a series of commercial nuclei, at each of which there are not only purchasers but also services and supplies of consumer goods. A network of service centres is created whose spacing depends on the productivity of each hinterland and the numbers of farmers the dealers can serve efficiently before setting up a branch 'office'. Temporary pick-up points can soon become branch offices. The biggest difficulty is almost certainly starting the system, for the risks involved in estimating the potential of a market for a new area of production, even of a crop for which there is an established demand, are considerable. The easiest situation is one where established dealers at ports can purchase initially small quantities of the new produce and ship them together with other goods. In such cases production near to port has obvious advantages and the coastward orientation of most export crop producing areas is an important result, although clearly helped by the major interest in the produce of tropical rainforest environments, which also in part have a coastward orientation. Generally both dealers and farmers need to believe that there is considerable potential both in the market and in the productive capacity of the hinterland to make the initial steps in developing commercial production worth while. Isolated areas of low agricultural productivity are unattractive to traders and tend to perpetuate themselves. Areas already developed may experiment with new crops, although such experimentation is often limited, and the tendency to locational specialization is marked. Often a sharp contrast develops in the Third World between areas of commercial agricultural innovation and areas in which traditional methods and crops are perpetuated, not through any innate conservatism but through the difficulties of finding a crop and a market large enough to make the initial risk-taking worth while.

In some areas of innovation farmers overreach themselves in the urge to develop new enterprises. Thus G. L. Johnson (1968) noted a tendency to overinvestment in Nigerian agriculture by peasant farmers looking for favourable opportunities, whereas E. A. J. Johnson (1965) observed agricultural isolation in northern India reflecting the paucity of towns. He argued that regional policy should seek to reduce such isolation by creating a spatial system of urban centres arranged in a hierarchy, i.e. a strategy for agricultural development based on expanding the urban system (E. A. J. Johnson, 1970). Isolation encourages the persistence of subsistence economies, or rather of economies in which peasant families produce all or most of their food requirements and even other vegetable requirements such as fibres, some building materials, tools, utensils and feeding stuffs for livestock. Pure subsistence is virtually unknown, for its discovery results inevitably in its disappearance. Most commonly a small surplus of production, either of local foodstuffs or of some specialized commodity, is produced for sale and the family income is supplemented by gathered produce, by additional occupations such as fishing, handicrafts or trading, and by money sent home by members of the family who have found work elsewhere. It has been estimated that in tropical Africa and India as much as 60–80 per cent of the agricultural labour force is engaged in subsistence production (Stavenhagen, 1964). Subsistence production discourages innovation for there can be little interest in new crops or expanded production except in so far as local or family food preferences may change and local population may increase. New crops are more readily adopted where they are analogues for existing crops, as, for example, the groundnut, which served as an analogue in West Africa for the earth pea (*Voandzeia geocarpa*) (Morgan and Pugh, 1969, p. 635). Subsistence, however, offers a virtually risk free economic situation in which demand is constant or only slowly expanding, but in which failure in one aspect of the system of production may mean starvation in a situation of limited or virtually non-existent local market and transport systems. In practice most systems are mixed. Peasant farmers are frequently reluctant to give up the economic 'safety' of producing their own food, but hopefully have planted commercial crops. Even plantation systems frequently include an element of subsistence food production, if only in permitting estate workers to cultivate their own small food crop plots in their spare time.

Among the many features held to be peculiar to or especially important in less developed economies, rapid increase of population is often regarded as one of the most significant, especially characteristic of the Third World and with a special relationship to agriculture (Fig. 8). Population density as such is not a general Third World factor, but a differentiating feature due to the considerable differences in density to be found in the less developed countries (Fig. 9). The contrast between the high population densities of the oriental Third World and the low densities of the remainder is a fundamental feature (Linton, 1961). In so far, however, as population density may be shown to relate to agricultural practice (see pp. 228, 231–2), so changes in population density which result from increases in population total may be shown to have affected agriculture. Such changes in density may be part of a localized population increase in a larger more stable population due to internal changes in the pattern of settlement, often associated with commercial or political changes. The resultant agricultural changes, particularly where they are associated with increased intensity, may in such a case be wrongly ascribed to the advance of commercialism. Increasing population is associated with areal expansion of agricultural settlement, still a feature of less developed economies, just as contraction or the reduction of cultivated area, increased intensity of agricultural land use and urbaniza-

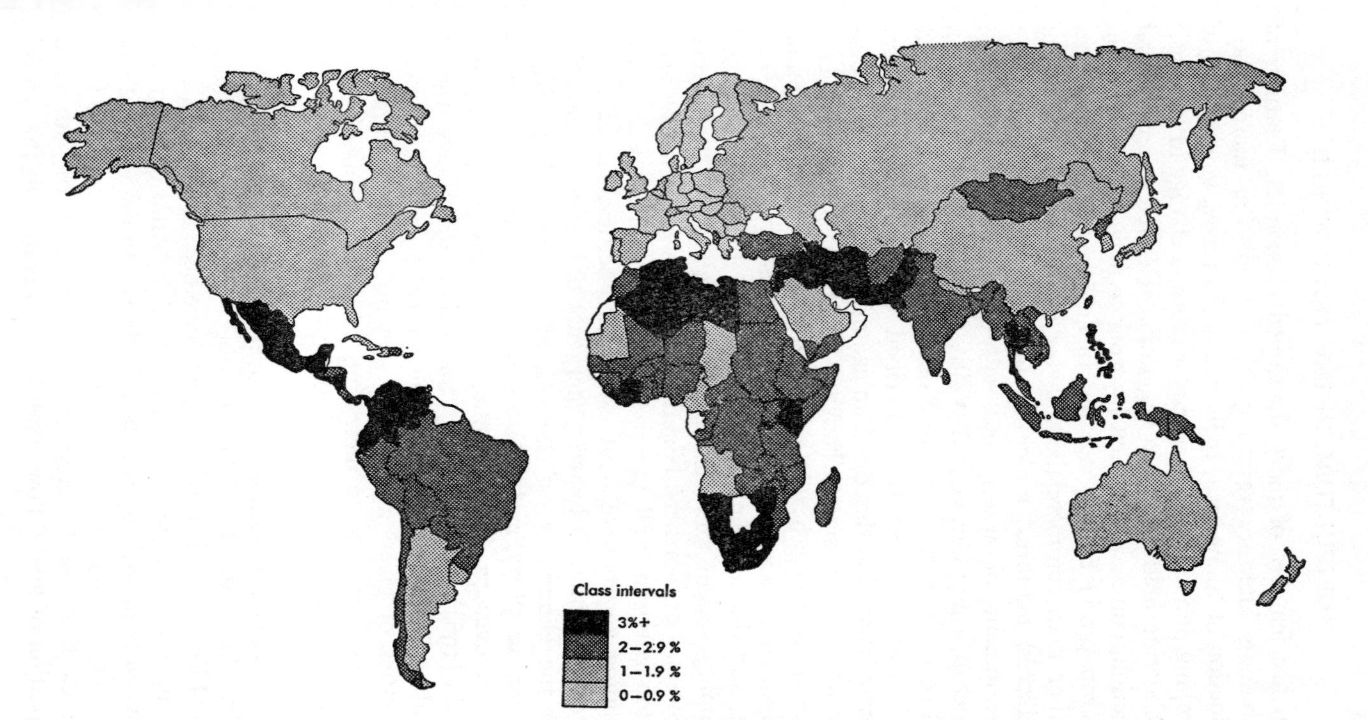

Fig. 8. Annual population increase 1965–72

tion are features of more developed economies. Features of increasing intensity resulting from population increase are reduction of fallow area, reduction of field sizes, lengthening of cropping period, changes in major crops, a change in emphasis in choosing field sites from vegetation selection criteria to soil selection, commercialization in order to achieve specialization of crop production in relation to environmental constraints (rather than commercialization for its own sake), the use of fertilizers, manures, mulching and green manures, and the development of water and soil management techniques. (Boserup, 1965; Gleave and White, 1969; Morgan, 1953, 1955 and 1959). The transition from shifting agriculture to floodland cultivation has also been ascribed to population pressure. Seavoy (1973) has described an example on the island of Kalimantan in Indonesia. In Tonga, Maude (1970) has shown how population growth has led to shorter fallows, a longer cropping period, changes in the food crop pattern and in fallow vegetation, and the complete areal merging of the bush fallow food cropping system and of coconut cultivation for commercial purposes. For Northern Nigeria, Grove has suggested an association between land use patterns and population densities, including zonation of land use around settlements at 19–58 persons per square kilometre, mergence of land use zones or at least the elimination of 'the outermost zone of bush fallow farming' at 58–77, and adjustment of the system to soil variations at over 77 (Grove, 1961, 122–6, and see pp. 236–9). Béguin (1964), Conklin (1957) and Allan (1965) have tried to calculate formulae to express the critical population carrying capacities of given agricultural systems. The adjustment of agricultural practice to increasing density of population is a major thesis of Boserup (1965), who has tried to show that increasing pressure of population on land resources has been a major cause of agricultural improvement, more especially, one might add, in the Third World, where the commercial incentives of Europe and North America have been so much weaker. Boserup also criticized what she called 'a static geographical theory of land-use', by which she referred to the views of van Baren (1960), regarding overpopulation in Java, and to ideas such as those of Gourou (1961), regarding the supposed adaptation of long fallow systems of cultivation in the tropics

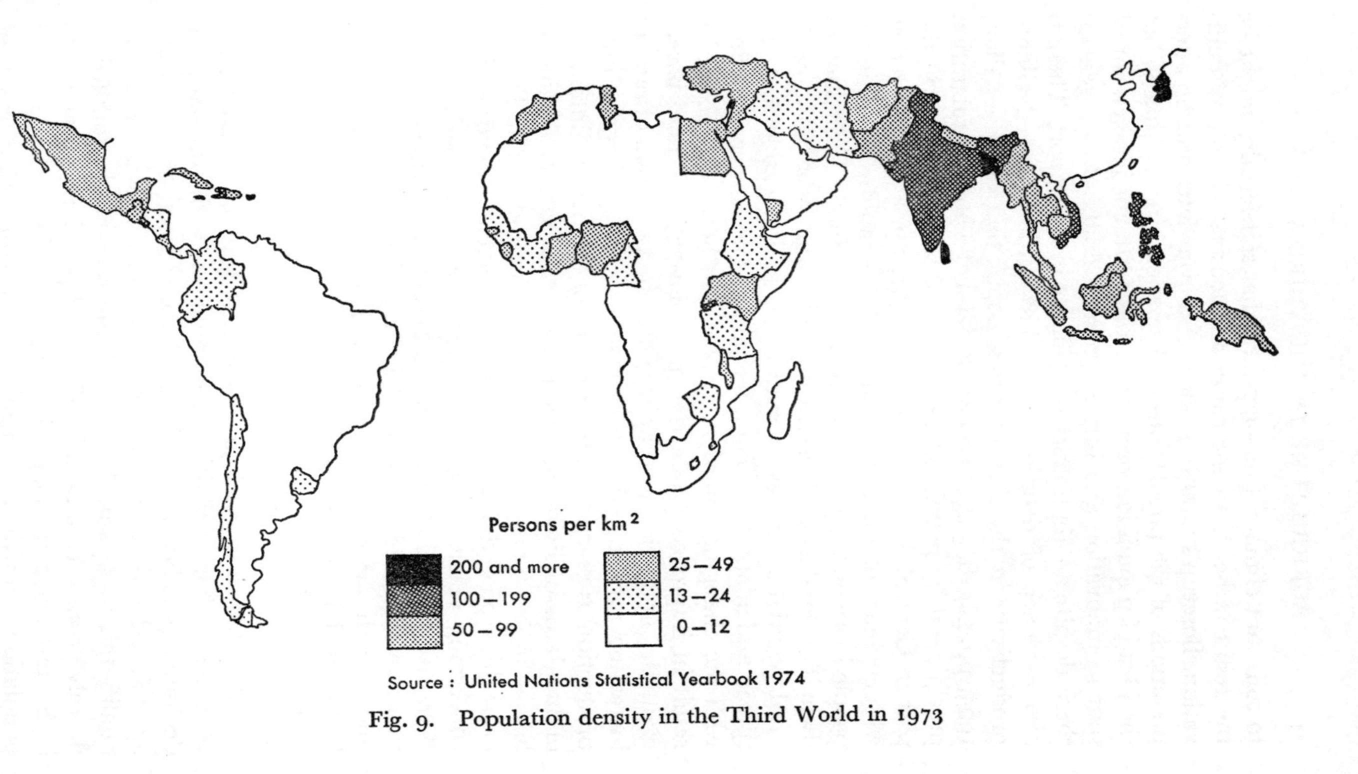

Fig. 9. Population density in the Third World in 1973

to soils and climate. She suggested that soil fertility might be the result rather than the cause of intensive methods of cultivation. Boserup's attack on static and over-deterministic interpretations of the population–land relationship was timely, but the idea of population pressure acting as an encouragement or spur to innovation and improvement is doubtful as a general thesis despite an abundance of evidence in its support. There is also evidence of deteriorating soil conditions and declining productivity in the face of rising population density (Allan, 1965, pp. 385–8, 454), and even a claim of wholesale migration precipitated by population pressure and soil exhaustion in the past in Central America (Gourou, 1961, pp. 43–51), together with evidence of inability to cope with the rapidly rising tide of people in several regions of contemporary India. Perception of dwindling land and deteriorating soil resources as an agricultural population increases may result in innovation and the spread and adaptation of new techniques and new crops to deal with the changing situation. It does not, however, necessarily result in improved production per man-day and may barely maintain production per capita. Often in a situation of rapidly increasing rural population increased intensity of agricultural production as described above results in the bare maintenance of standards of living or merely slows the rate of decline as labour inputs increase and returns per man-day fall. To describe this kind of development as improvement is to put notions concerning efficient husbandry before economic well being. Boserup's suggestions were that 'a period of sustained population growth would first have the effect of lowering output per man-hour in agriculture, but in the long run the effect might be to raise labour productivity in other activities and eventually to raise output per man-hour also in agriculture' and that 'primitive communities with sustained population growth have a better chance to get into a process of genuine economic development than primitive communities with stagnant or declining population, provided of course, that the necessary agricultural investments are undertaken'. She was careful to qualify this thesis with 'This condition may not be fulfilled in densely peopled communities if rates of population growth are high', but made no suggestion regarding the critical rates of population growth nor demonstrated convincing cases of 'genuine economic development', with rising outputs per man-hour resulting from 'sustained population growth' (Boserup, 1965, p. 118).

In strong contrast is Geertz's thesis of agricultural involution based on a comparison of agricultural change and population increase in Japan and Java. Geertz tried to show the development in Java of a dual economy in which a Dutch organized commercial estates sector became integrated into a modernized Dutch economy and more and more segregated from a 'rigidifying' Javanese economy. Of the export crops only sugar cane had an important relationship to the cultivation of the staple food crop, rice, since both were grown mainly by irrigation and sugar cultivation encouraged improvements in irrigation methods. Nevertheless the adoption of elaborate cultivation techniques, the intensification of the '*sawah* ecosystem' of rice production as population pressure increased, did not result in economic advance but only a nearly constant output per head as yields per unit area rose, supporting in effect, apart from the upsets of civil strife, a static economy and a burgeoning population. Local famine from crop failure has been prevented only by the expansion of the transport network (Geertz, 1968).

The Third World is heterogeneous in its social conditions and racial composition. Some portions of it, notably the Caribbean world, tropical Africa and Southeast Asia, are remarkably so and have proved extremely fertile grounds for the development of social and anthropological studies. The Third World societies offer an abundance of material for anthropologists or political scientists who are prepared to escape from the narrow limits of the Western world (Balandier, 1972). Such heterogeneity has encouraged the formulation of the plural society thesis, first expounded by Furnivall (1945 and 1948) but since applied elsewhere, notably in the West Indies (Smith, 1965). The plural society is not only mixed but depends for its existence mainly on external factors. Its social groups are linked economically but are socially divided, resulting in considerable restriction of social mobility and specialization in economic and other activities by social group, even by race. Whatever the virtues or otherwise of the thesis, the existence of complex societies with varied social groups practising different kinds of agriculture, having different staple crop preferences, different work norms, different attitudes to innovation and sometimes practising different job specialisms so that peoples from different groups may undertake different specialist functions on a single large farm, has important implications for agriculture.

The problems of pluralism in the Third World are by no means universal, but where they do occur have spatial significance for agriculture. Traditions and formal institutions act as filters controlling the importance and quality of the progress each group can make (Sternberg, 1967). Thus one may contrast the very different agricultural systems of Fijian and Indian immigrant farmers in the Sigatoka Valley of Fiji (Chandra, Boer and Evenson, 1974), the cropping preferences, holding sizes and enterprise of farmers of Negro and East Indian origin in Guyana (Burrough, 1973) or the farming activities of Malay, Chinese and Indian communities in Malaya (Dobby, 1955; Voon Phin Keong, 1967). Another view sees the many societies of the Third World as sharing certain characteristics, notably a tendency to give a higher priority to social than to economic gain. This somewhat controversial view, expressed notably by Hoselitz (1964) in the concept of the 'folk society', a concept which owes its origin to Redfield (1941 A and B), has a special significance for agriculture and the problems of agricultural development. The tendency to put social obligation before economic improvement has already had comment, but this frequently represents the minimum social obligation which would occur in almost any society, irrespective of its wealth, but which happens to be very expensive in relation to low incomes. In some societies spectacular expenditure on feasts and entertainment is the main path to social advancement and, in consequence, saving for economic improvement is virtually excluded. Moreover, many of the social theories explaining economic backwardness have a somewhat teleological character and have often been the product of a lack of economic information to match the 'wealth' of social data from anthropological sources. Some societies seem to have been static and generally resistant to change, reinforcing a peasant-farmer conservatism created in part by the high risk element in low productivity systems dependent on a capricious environment. The idea that most Third World societies are static, traditionally based structures (Hertzler, 1956) only recently modernized, so that a sort of dualism exists between a socially oriented backward sector and an economically oriented progressive sector, is a dangerously simple generalization. It cannot do justice to the complexity of agricultural change, which in many Third World countries has involved so-called 'modern' features for very long periods, in some cases well before contact with Western Europe. Frank has denied the existence of a dual economy, claiming that 'under-development' is the product of development and that backward regions become in effect satellites of metropolitan centres (Frank, 1966). Thus it can be argued that much backwardness is due to modernization, that the dual economy, if it exists, has been created by one process (Brookfield, 1975, pp. 53–84), resulting in industrial decline and a return to agriculture dominated by a new trading framework. One might add further that 'backwardness' is in any case a relative term and that 'underdevelopment' did not exist until development arrived. Franklin (1962) has suggested that the 'agriculturalization' of the peasantry or increased dependence on agriculture following 'development' has been a factor producing crisis in peasant societies. 'The rural community in underdeveloped countries has long ago ceased to be a closed world. It takes part in the market system . . . The rural community becomes at times the locus of grass-root political movements which are tied to the great social issues of our time . . .' (Stavenhagen, 1964).

AGRICULTURAL DEVELOPMENT: SOME HYPOTHESES

Everywhere agriculture is changing. In more remote parts of the Third World the response may be slow, but nevertheless changes may be seen, often unfortunately possessing a character one would hardly recognize as improvement. In agricultural geography the recognition and the understanding of the spatial implications of change are essential. The distribution map is a picture solely of a brief time-period whose data are the product of varying rates and even varying directions of change as farmers begin to move towards perhaps only partly understood goals, and as many of them find new goals before they have reached their previous choice. The map represents an agricultural flux. At best one can interpret it only by a form of mental spatial regression in which a great deal of noise affects our interpretation. Nevertheless the effort has to be made, as no static interpretation can even approximate to the truth or convey a reasonable idea of the processes involved. An account of the changes currently affecting agriculture in the Third World will be given in the survey to come, but first the general character of these changes needs to be established. Since the nature of the processes and their effects are the subject of considerable debate, a brief commentary will be made on the various hypotheses, usually claimed as theories or models, that have been suggested. These divide into two groups, those in which the ideas concerning agricultural changes are derived from a more general hypothesis of economic development, and those, usually more recent theses, which are directly concerned with the development of agriculture as such (for a detailed review see Hayami and Ruttan, 1971, pp. 9–63). The former tend to emphasize structural aspects, more especially shifts between sectors and regions during the development process, whilst the latter tend to emphasize agricultural technology, innovation and marketing.

General economic development theories which have had some special concern with agriculture have mostly been formulated as either growth stage models or dualism models. The former have been concerned either with the growth of agricultural production as a precondition for the development of manufacture (Marx, 1889, pp. 769–74; Rostow, 1956 and 1960; Kuznets, 1959; Nicholls, 1963) or with domestic industrial development as the chief generator of or encouragement for agricultural progress (List, 1885; Chenery, 1955, and numerous standard texts not all preoccupied with growth stages, e.g. Samuelson, 1970, pp. 749–50). List's stages included savage, pastoral, agricultural, agricultural-manufacturing and agricultural-manufacturing-commercial whilst Rostow's included traditional society, pre-conditions for take-off, take-off, drive to maturity and age of high mass consumption. Refinements subdivided these, creating stages within agriculture, for example Bath's (1963) self-sufficiency, partial self-sufficiency or direct agricultural consumption, indirect agricultural consumption or self-sufficiency for only part of the population, which itself divided into a phase with agricultural production as the chief source of livelihood for most male workers followed by a phase when agricultural production was no longer so. Most stage models are empirical and reflect the past rather than predict the future. They tend to be based mainly on observation of development in Europe and North America and have little relevance for the Third World. Of more interest are the models of Fisher (1945) and Clark (1940), which stressed the structural transformations that took place with progress from one stage to another. Clark's approach, with its suggestion of labour transfer from sectors with low per capita output to sectors with high per capita output, emphasized the role of manufacturing industry, as did his contrast of industries having increasing returns, such as manufacture and transport, with industries subject to diminishing returns such as agriculture, mining, forestry and fishing (Clark, 1953). More balanced approaches came from Nurkse (1953) and Lewis (1955, pp. 141, 276–83), although Lewis did tend to see agriculture as a prime source of capital, food and labour for economic growth (1955, pp. 229–31).

Dualism or dual economy models, in part discussed above, contrast a lagging traditional sector with a growing modern sector. Sometimes, although not always, the contrast is between traditional agriculture and modern industry. These models are based much more on observation of Third World economies. Boeke's thesis of sociological dualism was derived from a study of the failure of Dutch colonial policy in Indonesia (Boeke, 1953). It claimed the need for a distinct approach wherever two social systems apparently co-existed but had only limited contact. The traditional sector was concerned mainly with social needs and suffered from a tendency to resource immobility, resulting in a failure to achieve growth through investment. In consequence, attempts to change the traditional agricultural sector were useless. Change could be achieved only through the modern sector, that is through industrialization. Within agriculture arguments of this kind have appeared in the form of a contrast between a backward peasant sector and a modern plantation sector and have been used to justify, for example, the alienation of land to Europeans in tropical Africa or the post 1945 investment in large-scale mechanized systems of groundnut production in East and West Africa. Higgins (1959) pointed to a technology contrast between modern and subsistence sectors, the modern sector importing its technology and concentrating on primary production. Its growth had little effect on the subsistence sector, which lacked savings. What has been called the 'factor-proportions' approach was tested by Baldwin (1966), more especially in an attempt to explain how the growth of mines and plantations in Northern Rhodesia, modern Zambia, failed to develop a 'spread effect' promoting growth in other sectors of the economy. Myint (1964) emphasized the problem of obtaining capital as a source for dualism by contrasting the accessibility to finance of industrialists with that of the rest of the population. The modern sector obtained capital more cheaply, leading to the adoption of a capital-intensive technology and high labour productivity. A low demand for labour relative to the traditional sector resulted and the latter also suffered from a failure in investment promotion. More recently Myint has advocated discrimination in favour of small-scale production units in the traditional rural sector in order to reduce distortions in the factor markets and alleviate internal economic disparities. He has suggested the raising of official rates of interest to levels sufficient to equate the demand and supply of loans in the economy as a whole and to stimulate domestic savings (Myint, 1972, pp. 40–1). What has been called 'dynamic dualism' was stimulated by Lewis (1954) and then developed by Jorgenson (1966 and 1969) and by Fei and Ranis (1964). The classical Fei and Ranis model involved a subsistence sector with underemployment, disguised unemployment, zero marginal productivity of labour, constant real wages and fixed land inputs. Labour could be transferred from the subsistence to the modern commercial and industrial sector without reducing agricultural output or increasing the supply price of labour. The loss of labour to industry even resulted in the production of an agricultural surplus which could be invested in commerce or industry. Jorgenson concluded that the classical approach did not fit the evidence, which supported rather the assumptions of the neo-classical models, i.e. that labour lost to agriculture did result in a loss of productivity, that real wages were not fixed but variable and that technological change was required in agriculture at an early stage in the growth process. Other criticisms have been made by Mabro (1971), who concluded from evidence for Egypt and India that agricultural wages were related to the marginal product of labour, that the labour market in agriculture was very active, that labour inputs per unit of land were higher on smaller holdings, that disguised unemployment was not a prevalent feature of agriculture in overpopulated countries, although some underemployment was. Mabro suggested that there was a need for institutional reform to effect agricultural improvement, together with both the break-up of large estates and the consolidation of small farms in co-operatives, although open unemployment might result amongst landless workers from more efficient use of family labour.

The increasing attention to the state of agriculture in the Third World, partly through greater realization of its enormous economic importance and higher hopes of its potential for improvement, has resulted in a considerable volume of theory concerned with change in agriculture itself. Some of this theory has tended to diverge from the models and theses of general economic change already examined, partly because the agricultural economists and other students of agricultural development have asked themselves different questions, partly because the part is necessarily different from the whole and partly because many of the ideas concerning agriculture have been even more empirically based and more susceptible to testing. Some of them are also strongly related to a newer group of ideas concerning economic development in general which are based on notions of unequal development between sectors and regions, polarity, central place systems, innovation and diffusion.

Johnston and Mellor (1961) whilst recognizing the essential role of agriculture in development in providing labour and capital for investment in industry, essential food supplies, exports for foreign exchange earnings and a market for industrial expansion, argued that whilst much of this could be achieved through more effective use of the resources already committed to the agricultural sector with only modest requirements of high opportunity cost resources, nevertheless there were compelling considerations for introducing modern technology on a broad front. The need for a technological revolution in Third World agriculture has the support of most modern theorists. Even those who would put industry, commerce or the development of an adequate infrastructure first, nevertheless would accept that agriculture is the largest employer of labour and the largest contributor to gross national product in the majority of less developed countries and that the possibilities for improved productivity are in most cases considerable. The most extreme arguments, as reported by Grigg (1970, p. xiii), generally claim that the main problem is simply to persuade most farmers to adopt the improved methods already practised by the minority, and that farming methods of the temperate regions can be transferred unmodified to the tropics. Arguments are mainly concerned with how the possibilities are to be realized, not with whether the attempt should be made. In particular they focus on whether there is room for a more effective use of resources, as suggested by Johnston and Mellor, or whether the current techniques are at the limit of their efficiency so that improvement requires in effect an agricultural revolution.

Interest has shifted therefore to techniques, incentives, research and the supply of information and training. Although the main thrust of these developments has taken place in the last thirty years, nevertheless attempts to do more than introduce an enclave of modernization, such as foreign-owned and operated plantations, go back at least to the early years of the last century with the efforts of various agencies to introduce new crops and techniques, not only to provide raw materials for the industries of the West, but also to increase food production to offset the shortages created even by the small developments which were then taking place. As early as about 1772 Joseph Banks envisaged Kew as 'a great exchange house of the empire, where possibilities of acclimatizing plants might be tested' (Allan, 1967; Cameron, 1952; Masefield, 1972, pp. 20–1), and in 1787 Captain Bligh was commissioned to carry bread fruit plants from Tahiti to the West Indies in the naval vessel *Bounty*. Even the enclaves could not remain isolated. Settlers in the Third World introduced new crops and techniques often in order to promote general economic well-being, but sometimes unintentionally as in the introduction of cocoa pods and of planting techniques from the plantations of Fernando Po to Ghana, then the Gold Coast colony (Hill, 1963, pp. 172–3). The nineteenth century saw the establishment of forestry and agricultural departments in the European colonies, botanic gardens, experimental farms and the production of new strains of crop plants. It also saw the development of considerable initiative and enterprise amongst peasant farmers who invested in the new farming and in many cases provided labour and money to build roads and erect the warehouses needed for commercial expansion. Most work on diffusion models of agricultural development has focused on Europe or North America although some work has begun on the Third World, more especially in studies of peasant agriculture. As yet the Third World lacks its Hagerstrand (1953), Wolpert (1964) or Griliches (1960). Much more work has been done on farm management research and the development of quantitative techniques derived from production economics (Heady, 1952; Heady and Dillon, 1961). Many of these techniques have been adapted for use in advising peasant farmers on the development of new systems of production and there are few countries in the Third World which are without basic manuals containing production standards and outlines of model agricultural systems for the production of at least some of the more important commercial crops. The idea, however, that the transfer of farm management techniques from the Western world, allied with a knowledge of diffusion processes, would solve the problems of transforming peasant agriculture met little support in practice and was opposed eventually on the grounds of both observation and theory by Schultz.

Schultz's book, *Transforming Traditional Agriculture* (1964), was a landmark in the development of hypotheses concerning agricultural change in the Third World. Schultz, following studies made elsewhere, notably in India, insisted that agricultural technology was essentially location specific, i.e. it could not easily be transferred from one environment to another, and that the production gains to be made by the re-allocation of resources in peasant agriculture were only limited since for the most part peasant farmers were efficient in the allocation of resources. Part of the evidence for this argument came from Tax's detailed study (Tax, 1953) at Panajachel in Guatemala, an area which Watters has since suggested is exceptional. Watters has disputed Schultz's view and argued that allocation efficiency varied immensely with different cultures (Watters, 1967). Schultz also argued that poor agricultural communities do have enough competent entrepreneurs to do a satisfactory job in using the factors at hand, that peasant farmers do adjust to changes in relative prices of products and factors, that no part of the agricultural workforce has a marginal productivity of zero and that there are few significant inefficiencies in factor allocation. Agricultural poverty was seen as a product of limited technical and economic opportunities. In consequence economic growth depended upon the availability and price of modern agricultural factors, which sounds like another form of dualism but in fact is not, because in this case the modern factors are not necessarily the product of another society nor are they to be used in a distinct progressive sector or enclave. Schultz went on to make the point that modern inputs were seldom ready-made and could rarely be introduced into peasant farming in their current form. Most readily available was a body of useful knowledge which could be used to develop factors appropriate to the environments of poor communities. Profitability was the chief factor deciding whether farmers would adopt a new crop or technique, and here an important element was whether there was a large market with a highly elastic demand. In the Third World this usually meant a foreign market. What has been called the 'high-pay-off' input model has had support from the initial success achieved in the Green Revolution, where high rates of return were achieved by the cultivation of new high-yielding grains (Hayami and Ruttan, 1971, pp. 40–2; Brown, 1970, pp. 41–3).

Many hypotheses appear to have been developed in a vacuum. Agriculture is a considerable user of space and its performance normally reflects the varying character of space and environment. It is with that variation that this book is largely concerned. Theories of agricultural development incorporating a spatial element to a significant degree originated, as was shown at the beginning of this chapter, with Thünen In 1953 Schultz outlined the implications of urban-industrial development for agricultural development, which were that economic development took place in a specific locational matrix, which was primarily urban-industrial in character, and within which the economic organization worked best near the centre (Schultz, 1953). This thesis clearly applied mainly to developed countries but could be applied to the kinds of transformation of agriculture taking place in the Third World, wherever urban-industrial growth was taking place. It was in fact examined in the context of urban-industrial development in São Paulo, Brazil, by Nicholls (1969). The increased realization of the importance of urban growth not only as a market for goods but as a source of inputs, information and even direction has encouraged a large number of studies which have attempted to explore the relationships between town and country in the Third World, have examined rural migration to the towns, the growth of agricultural processing industries, rural periodic markets and the operations of dealers (e.g. Lipton, 1968; Preston, 1972 and 1973; Dutt, 1972) and have sought some illumination from growth pole and central place theory. Growth pole 'theory' was initiated by Perroux (1955) more as a collection of observations correcting earlier models than as a coherent theory. Perroux suggested that economic growth was manifested in points or 'poles' of variable intensity, particularly in underdeveloped countries where poles tended to be geographically and economically isolated in economies 'not yet articulated through networks of prices, flows and expectations'. Poles were characterized by industrial complexes with key (propellent) industries, having the property of increasing the sales and service purchases of other industries. They also had spread and backwash effects (Myrdal, 1957, pp. 27–32) which could affect the surrounding rural area. These ideas have been tested in Malaysia (Robinson and Salih, 1971) and have been developed further by Friedmann in a general thesis of regional development based on polarization and involving his 'centre-periphery' model (Friedmann, 1966 and 1972; Friedmann and Alonso, 1964, pp. 1–13). The growth-pole hypothesis has attracted a vast number of adherents and has been incorporated into numerous planning proposals. It has also been criticized as lacking clarity and in its most recent manifestations as lacking both original insights and theory structure (Brookfield, 1975, pp. 105–8). More useful are: 1. Consideration of central-place theory, or the idea of a service-providing settlement hierarchy, in the understanding of the development of the urban networks and transport systems that play so vital a role in agricultural change. 2. Consideration of the problems involved in the concentration of investment into selected regions resulting in unequal development.

It has been argued that geographical variation combined with limited resources will inevitably create unequal development between regions—'in the geographical sense, growth is necessarily unbalanced' (Hirschman, 1958, p. 184). Friedmann thought that ultimately there would be regional convergence or at least that the national economy would 'appear as a fully integrated hierarchy of functional areas', although later he admitted the possibility of persistent inequality (Friedmann, 1966, pp. 14–18). We do not have to accept a natural law of regional growth patterns, although we may accept that attempts to avoid all inequality or concentrations are likely to prove extremely costly. The development of methods of assessing the likely pay-off in terms of both social and economic advantages of alternative regional strategies is essential.

The hypotheses so far presented each deal only with part of the complex of processes affecting development. The attempt to develop an all embracing model or general synthesis of the various hypotheses concerning agriculture has been made by Hayami and Ruttan whose 'induced development model' incorporated the high-pay-off input model together with hypotheses concerning the urban industrial impact, the diffusion of innovation and the conservation of natural resources (Hayami and Ruttan, 1971). It treats technical change as endogenous to the development process, and not as some external independent force, and recognizes that there are several paths of technological development, including labour-saving mechanical technology and land-saving biological and chemical technology, including land and water-resource development. Relative factor and product prices have a pervasive impact on the innovative and productive activity of farmers and of firms supplying industrial inputs for farm production. The efficiency of the pricing system is therefore vital, and any distortion in price relationships through market imperfections or government intervention causes distortion of innovative and productive behaviour. Market structure reform is a prerequisite for progress. Other prerequisites are land tenure reform, the development of credit institutions and the creation of an adequate infrastructure, including extension services and education. Research and education institutions respond to economic forces by releasing the constraints on agricultural development imposed by inelastic factor supplies. The model is examined mainly in relation to agricultural development in Japan and the United States, but it is also 'tested' by developments in the Third World, more especially by the events of the Green Revolution. Its main limitations lie in the little exploration of capital-saving techniques—a scarce resource in the Third World—as, for example, in an emphasis on variable rather than fixed inputs, and the very little exploration of the spatial implications.

2

Agricultural production and systems in the Third World as a whole

The crucial factor in the pace of agricultural development in the Third World is the dominance of the more developed countries in the world economy. The more developed countries or 'developed market economies' have less than a third of the world's population producing over 80 per cent of the world's estimated gross domestic product. They probably contain about 12 per cent of the world's farm population producing some two-thirds of the world's agricultural output and two-thirds of the world's gross agricultural exports by value (FAO, 1972). They therefore dominate world agricultural production and trade just as, even more powerfully, they dominate industrial activity. Agricultural production appears to be increasing at a faster rate in the less developed countries than in the more developed, and faster still in the communist countries or 'centrally planned economies' (Fig. 10). The rising agricultural production of the less developed countries has, however, hardly kept pace with their enormous increase in population and has been achieved mainly by the application of more labour to more land, so that yields per unit area, with certain important exceptions, mainly in southeast Asia, and returns per unit of labour expended have changed very little. Agriculture in the Third World in the last decade has overall been not much better than stagnant, with the sole major exception of the much

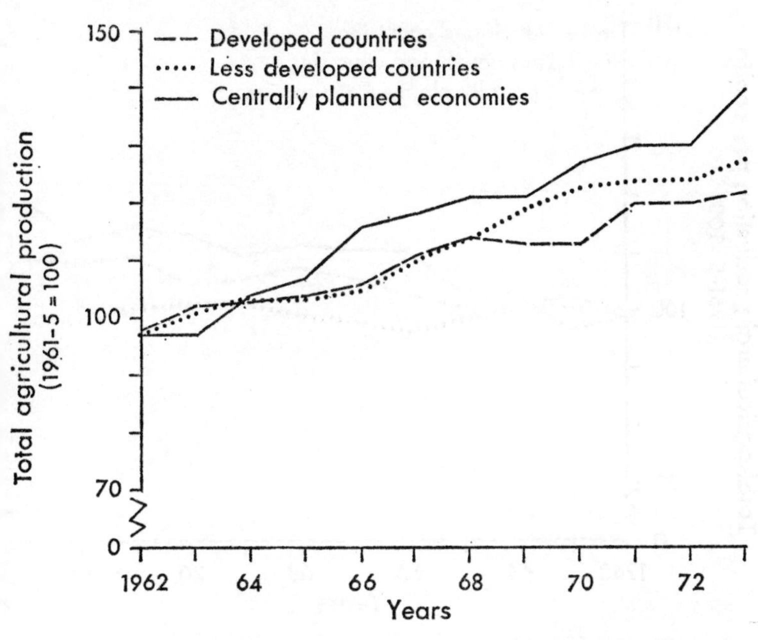

Fig. 10. FAO indices of total agricultural production (1961–5 = 100)

debated Green Revolution. Even before the Green Revolution there were some yield increases, notably in grains, but even though some 80 per cent of the expansion of the major crop area in the world was in the less developed countries in the period 1950–65, nevertheless the per capita area of agricultural land in the Third World fell (FAO, 1966; Grigg, 1970). In the more developed countries both agricultural production per capita (Fig. 11) and yields per hectare (Fig. 12) have increased, on the whole more than in the less developed countries. In the centrally planned economies, although output per capita has risen even more rapidly, rising incomes have created a growing demand which, in years with a small drop in production per capita such as 1969 and 1972, has even resulted in food shortages.

The population of the Third World has been increasing at an estimated 2·6 per cent per year, possibly even more in the current decade as the rate has been tending to rise. Its agricultural population is some 61 per cent of total, and industrial and

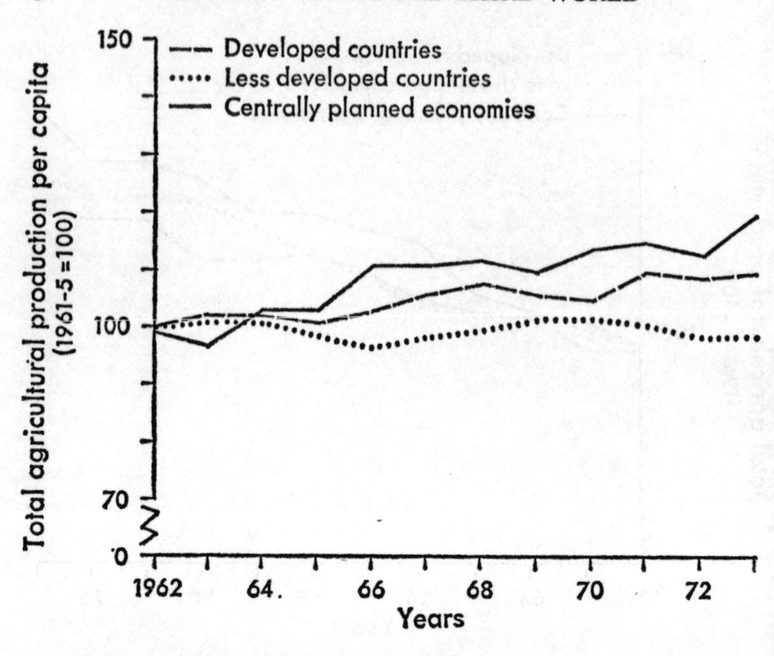

Fig. 11. FAO indices of total agricultural production per capita (1961–5 = 100)

service occupations are not increasing at a fast enough rate to effect more than a slow reduction in that figure, or possibly even to hold it, except by the creation of a vast number of urban unemployed. A growth rate in agricultural production of some 4 per cent a year is needed to cope with the needs of increasing population, the hope of some rise in real incomes and increased demand for agricultural produce (FAO, 1974A; Ojala, 1969). This growth rate is the basis of FAO's provisional Indicative World Plan for Agricultural Development (IWP) of 1970, whose targets have so far been achieved by only 4 (Brazil, Colombia, Egypt, Ethiopia) of the 16 largest developing countries (India, Indonesia, Brazil, Bangladesh, Nigeria, Pakistan, Mexico, Philippines, Thailand, Egypt, South Korea, Iran, Burma, Ethiopia, Argentina, Colombia, i.e. with populations of over 20 millions), which together account for nearly three-quarters of the population and agricultural production of the Third World (FAO, 1974A, p. 22).

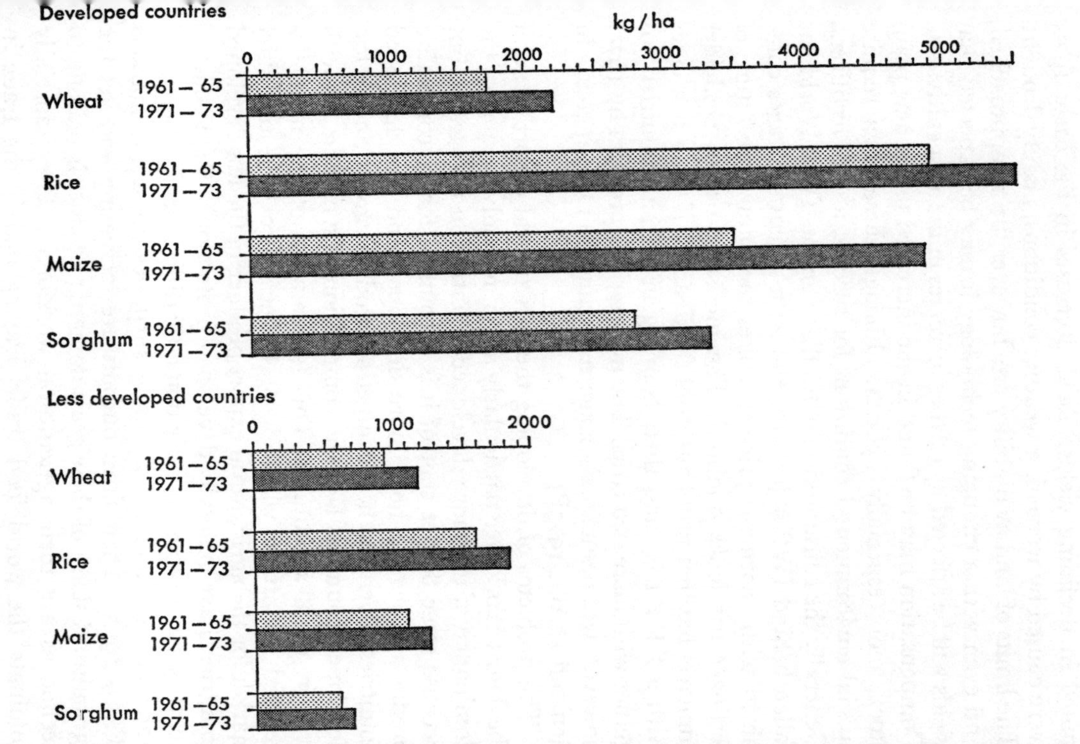

Fig. 12. Increases in grain yields (1961–5 to 1971–3) in the developed and less developed countries in kg/ha

So far areal expansion of agriculture in the Third World, which must eventually involve less suitable land, has not resulted in declining yields. Yield decreases in the early 1970s were caused by worsening weather conditions, mainly drought. The limits of land availability are, however, being approached, and even with a changing technology it may be expected that yields will be affected by further increases in the area cultivated. Compensation may be found in the increased use of new plant stock, more especially hybrids, although these often require special environmental conditions for success, and of fertilizers. Generally the consumption of fertilizers in the Third World has fallen behind IWP objectives, and with recent shortages combined with enormous price increases the immediate future of fertilizer use looks doubtful. The increased use of pesticides similarly has met problems from the dangers of using pesticides with a DDT base and their banning in certain countries, together with increased costs. The most serious problem by 1975, however, was pesticide shortage reflecting too rapid increase in demand (FAO, 1975B).

The problems of developing the agricultural export trade of the Third World are particularly acute, not only because of the dominance of the more developed countries in that trade, but because some 80 per cent of it is accounted for by competing products grown in both more developed and less developed countries. Even the non-competing products grown in the less developed countries face the competition of synthetics, as with rubber, or of substitutes. The governments of the more developed countries, which have hitherto provided the chief market for the agricultural export produce of the less developed countries, have generally developed protectionist farm policies, i.e. they have exported their domestic difficulties by increasing trade barriers and by dumping surpluses on world markets (FAO, 1972). Their farm outputs are increasing and yet the agricultural share of their gross domestic product is tending to decline as are their agricultural workforces. They especially dominate the world food trade, leaving mainly the trade in beverages, bananas, rubber and hard fibres to the Third World. In the effort to increase their exports the less developed countries have diverted resources into the production of especially favoured export crops, and this, together with increased expenditure on industrial and commercial expansion and the enormous demands of creating new infrastructures, has resulted in the general failure of food production to keep pace with demand. In the last two decades the less developed countries have become net importers of grain more especially of wheat and rice whilst the more developed countries have become net exporters. Recent trends are shown in Fig. 13. Urban growth has been regarded as a major cause, as subsistence producers

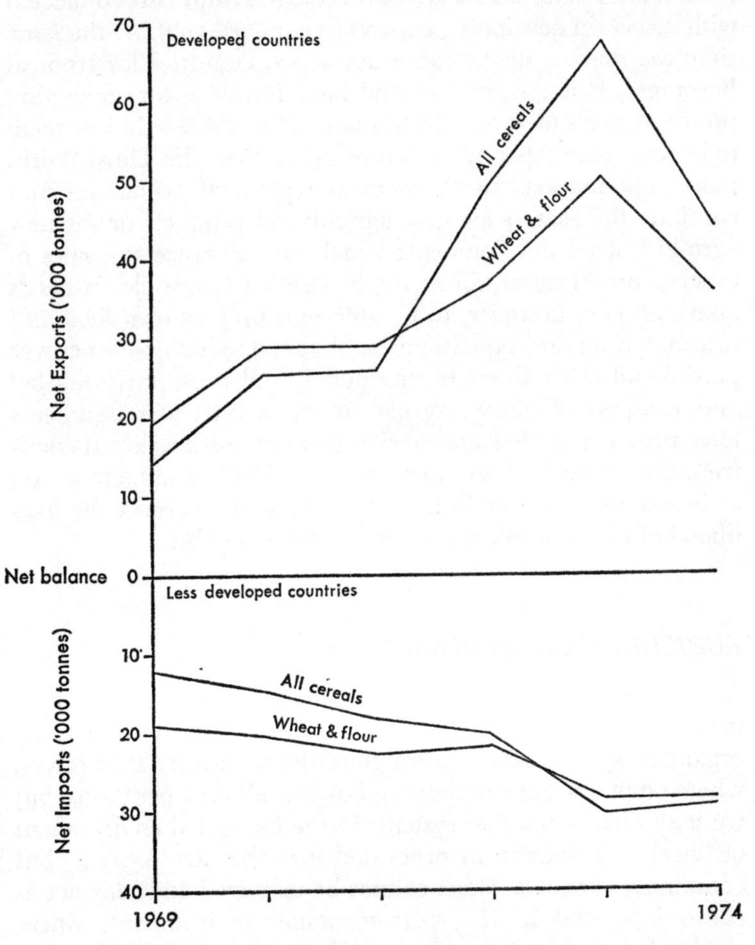

Fig. 13. World grain trade 1969–74: a comparison of recent trends in the developed and less developed countries

have had difficulty in becoming commercial farmers, but frequently commercial farmers have also become consumers of imported foods, as, for example, the groundnut farmers in Senegal or the rubber growers in Liberia. In some countries population pressure may have been a problem, although, as Boserup (1974) shows, it cannot be a sole explanation. She cites as factors policies in industrialized countries of farm income support, import restrictions for food and export promotion of food. Additional reasons are part ecological and part connected with success in developing exports (see pp. 93–106). At the same time the market in the more developed countries for tropical beverages, bananas, rubber and hard fibres, or for competing products, such as sugar, cotton and oil seeds, would not seem to have a great capacity for future expansion. The Third World must seek markets in the centrally planned economies and continue the search for new agricultural products or for new agro-industrial developments which will enhance the sales of existing products, as, for example, canned pineapple. It needs above all else, however, to be able to supply its own food and raw material requirements and to that end to increase wherever possible all other forms of production and to improve market and transport efficiency. Agriculture or industry first arguments have proved pointless and sterile. In most less developed countries, even though short of capital, an effort has to be made on as broad an economic front as possible and wherever the likelihood of an economic return on investment exists.

AGRICULTURAL SYSTEMS

Agricultural production results from the systematic arrangement of energy flows, usually organized. The most strictly organized systems in agriculture are the farm-firms themselves, where often one person plans and directs all the operations, but we may recognize other systems in the biological environment of farming, normally incorporated into the farm system, but containing elements which cannot be managed and may act as constraints, and in the socio-economic environment, where market forces and government actions are powerful influences. Much of our understanding of these systems is best achieved at other scales. For our present purpose generalizations and comparisons are necessary which inevitably provide no more than a crude picture of the overall mass of Third World farming operations. What we need is a general classification which can indicate the broad spectrum of agricultural operations, but which should not be used for any other purpose. The problems of general classification have been discussed elsewhere (Chisholm, 1964; Simpson, 1965; Harvey, 1969, pp. 326–7; Morgan and Munton, 1971, pp. 104–25). Farm systems are normally classified by their purpose, including size, type of management and labour force, and by their input and productivity. In recent years there has been a tendency to use labour input measures for classification, or even standard man-days where actual labour input measures are not available, as inputs have been regarded as a reflection of the intentions of the management with regard to the system, whilst outputs are biased by weather and other factors which may be beyond management's control (Jones, 1965).

In the Third World very little study has been made of agricultural systems as such, although the term is frequently used. Farm or cultivation types based on the observation of husbandry practice are the rule, or even mixed classifications in which the criteria are allowed to vary. The problem in classification reflects partly the limited nature of available data and partly the environment orientation of most studies, which have emphasized the tropics rather than the Third World and have tended to accept a tradition of concentrating on husbandry rather than on agricultural economy. Hence the universal recognition of types such as shifting cultivation, rotational bush fallow and swidden cultivation (e.g. Ruthenberg, 1971; Grigg, 1974; Manshard, 1974), their mapping in a complex hierarchy of criteria (Morgan, 1969A, pp. 252–3) and the compilation of reviews, readings and monumental collections of abstracts (Watters, 1960; Bartlett, 1955, 1957, 1961; Miracle, 1964), amounting in certain instances of preoccupation with detail almost to an obsession. Shifting cultivation has been the focus of a worldwide debate concerned partly with questions of conservation in relation to slash-and-burn methods of vegetation clearance, partly with its so-called 'backwardness', so that some have regarded it as a key criterion of 'underdevelopment', and partly with its supposed aping of the environment, so that the shifting cultivator may be seen as part of nature rather than apart from it. Shifting cultivation is without doubt an important topic in considering agriculture and environmental management in the Third World or in the tropics, but it is not an agricultural system nor the chief criterion of a system, any more than the use of a particular rotation would be regarded as the most important feature in classifying agricultural systems in Western Europe. It is an agricultural practice and an essential feature in understanding soil and labour management in certain Third World agricultural systems, but it does not define objectives and can appear as a component of a number of different systems, including commercial agriculture and systems of varying intensity, as, for example, the shifting methods practised in some plantations, although both 'types' frequently appear as distinct divisions of the same classification.

Accordingly, following the conventions of examining purpose, enterprises, structure, input and output, but limited by problems of data availability, we proceed to an analysis. Not all possible types of system will be examined, for the variations are almost infinite and some of the features are better studied at regional or local levels. It is suggested that a beginning may be made here with a broad division by purpose, type of management and broad enterprise and enterprise combination class. Such a general system might be as follows:

1. Commercial and subsistence
(Family farms, family and hired labour)
a. Single staple dominant
b. Two or more staples dominant
c. Mixed crops and livestock
d. Livestock

2. Mainly subsistence
(Family farms)
a. Single staple dominant
b. Two or more staples dominant
c. Mixed crops and livestock
d. Livestock

3. Mainly commercial
i. Family and hired labour
a. Single crop
b. Two or more crops
c. Livestock
ii. Management and hired labour
(Mainly single crop and some livestock systems.)
iii. Centrally managed rent or sharecropping smallholder systems.

Subdivisions may be made amongst the commercial and subsistence and mainly commercial systems by type of market, e.g. local, national or export, by whether the crop is processed before marketing or not, i.e. whether the system is entirely agricultural or agro-industrial, by intensity of operation in relation to land or other inputs, by kinds of enterprise, e.g. tree crops or field crops, food beverages, fibres, oilseeds, rubber, dairy produce, meat, animal fibres and hides, by level of technology, and finally by husbandry technique and environmental factors. In practice some of these broad system classes are very much more extensive than others and attention will be given mainly to those which are widespread and employ large numbers of people.

1. *Commercial and subsistence*

Systems which mix commercial objectives with the objective of providing most or all of the family's food needs from the farm, together in some instances with fibres such as cotton or wool and even medicines or the poisons used in hunting, are the most widespread in the Third World and probably employ the greatest part of the agricultural workforce. Such systems may have a dual character, that is they may identify their commercial and subsistence sectors with distinct crops or groups of crops located in distinct areas of the farm, or only partially dual, that is distinct crops mixed in the same area, or integrated in that the same crops serve both subsistence and commercial purposes. For this study only family managed farms of small size, rarely more than 50 hectares and most frequently less than 10, are included. Subsistence cropping in large plantations or centrally planned estates of smallholders usually takes only a very small part of the area and of the labour-time. Most family farms employ some hired labour, usually temporary, and mainly to ease the problem of some seasonal bottleneck in the use of labour. Normally the hiring of labour depends very much on the degree of commercial involvement of the farm, although such hired labour is not necessarily concentrated on export crops and can be occupied entirely in production for the local market. Family farms with hired labour are extremely efficient in their seasonal pattern of labour use, highly absorptive of labour in that their labour input per unit area is higher than that of large farms, flexible in their response to market changes in that they produce their own food and can hope to survive sudden falls in commercial crop prices, and sometimes highly productive per hectare. Often, however, they lack most of the resources which make for efficiency, have a very low productivity per hectare and per labourer, suffer from relative immobility of family labour, cannot innovate unless a very big increase in returns is estimated to offset the risks, and involve heavy costs wherever improvements such as the introduction of new seed or new techniques are contemplated. In Mexico, for example, about half of the crop and pasture area is in smallholdings or *ejidos* on lands expropriated from properties considered oversized. On these units, usually of poor technology, the main crops are subsistence maize and beans supplemented by cotton, henequen and coffee as the chief commercial crops, and giving yields on average a quarter less than those of large commercial farms. In much of Latin America, however, mixed commercial and subsistence smallholdings have only a small share of agricultural production, although they may be the majority in terms of number of holdings. For example, in Brazil 85 per cent of holdings are less than 100 hectares but they occupy only 27 per cent of the farmed area. Comparable figures for Argentina are 70 per cent and 3 per cent, for Cuba 92 per cent and 29 per cent and for Colombia 95 per cent and 30 per cent respectively. The main areas of occurrence are South and Southeast Asia, the Middle East, all of Third World Africa, southern Central America and some of the West Indian islands.

a. Single staple dominant

The most important area of single staple dominance is in South and Southeast Asia and is associated with the cultivation of wet-rice for family consumption. This cultivation supports the majority of farm families in the area and is concentrated in the lower Ganges plain and on the eastern and south-western coastlands of India, in Burma, Thailand, Khmer Republic, North and South Vietnam, Indonesia, especially Java, Borneo and the Philippines. Yields are generally high considering the limited resources available, at 1100 to 1800 kg/ha average in South and Southeast Asia (compare 4000–7000 averages in Egypt, Brazil and Australia). Generally the crop is associated with areas of high population density, even overcrowding, where small farms have to be highly productive to support the farm families. Water control is a vital feature of the system, demanding in its use of labour, often encouraging weed infestation. Sometimes it occurs more in the form of flood control than irrigation in the strict sense, making possible two or more crops a year and often creating a micro-environment in which few other crops can grow, hence encouraging a new monoculture in some areas such as the lower Mekong and Menam valleys, the lower Irrawaddy and the delta of the Ganges and Brahmaputra (Grigg, 1974, pp. 75–83). There are few wet-rice farming systems, however, which depend entirely on floodland or irrigated land. Most also have land dependent on rain with small plots producing fruits, vegetables and minor food and commercial crops. This variety of environment and agricultural practice also helps in maintaining a more even employment of labour throughout the year. In Java commercial sugar cane is planted in the wet-rice fields or *sawahs* and its cultivation has been claimed to improve the local ecological conditions for rice production (Geertz, 1968). Groundnuts, tobacco and pulses are also planted in the *sawahs* in rotation with rice, while cassava, sweet potatoes, maize, groundnuts and soya beans are planted in the *tegalans* or 'dry' fields. In Malaya commonly two wet-rice or *padi* crops are grown a year and the crop is both stored for farm family consumption and sold. Minor commercial crops include coconuts, sugar cane and a variety of vegetables. In the past small surpluses above subsistence requirements have been claimed as sold at uneconomically low prices because of the dominance of subsistence production, and in consequence gave a false impression of Southeast Asia as a source of cheap food (Dobby, 1955, p. 83).

For a time, Southeast Asia, more especially Burma, Thailand, and South Vietnam became exporters of rice, partly through the initiative of British and French colonial governments in providing major water control works and in introducing rice mills, partly through local initiative amongst dealers and partly through the existing concentration on wet-rice and in consequence, once water control was provided, through the ease with which the production of that crop could be expanded. Today the rice exports of the Third World have declined—in South Vietnam through the war, in Burma through a reaction against the pre-war emphasis on export expansion, coupled with the expropriation of absentee landowners especially the *chettyars* or southern Indian money-lending caste which had helped to finance the original rice expansion (Fisher, 1964, pp. 474–6). Only Thailand is still a major exporter and even it has seen a small decline encouraged by the 'rice premium' or export tax on rice intended as a counter-inflationary measure and an encouragement to the production of other export crops (Myint, 1972, pp. 83–4).

Single staple dominance in systems with an important commercial element is fairly widespread on land dependent on rain where small family farms often grow a variety of minor crops by varied cultivation methods. Single staple dominance commonly occurs in areas with limited or only recently developed commerce, where one particular foodstuff, most commonly rice, maize or plantain, is strongly preferred, or where there is some factor restricting choice of staple, commonly length of growing season, as, for example, in the preference for quick growing varieties of pennisetum millet on the northern margins of agriculture in tropical Africa, but sometimes soil or slope conditions or even the attention of pests, as in the preference in some areas of Niger and Northern Nigeria for awned varieties of pennisetum millet avoided by birds. Generally in smallholdings dependent on rains the use of a fixed family labour resource and virtual dependence on the food produced on one's own land suggest a preference for a variety of crops permitting a better spread of work load in addition to a more varied diet. There are generally advantages for the management of soils and farming operations in such a variety. Single staple dominance in the Third World tends to be the exception rather than the rule, an exception at its widest extent in South and Southeast Asia, where rice is often dominant not only as a wet-land crop but also as the product of shifting or rotational bush fallow methods of cultivation on rainlands. In many instances, however, single staple dominance is deceptive in its implications for agricultural practice because of local abundance of varieties grown, giving a wide range of timeliness in cultivations and of environmental preferences. Thus yam dominant farming systems in West Africa, more especially those of the moister southern savannas, where yam is largely a commercial crop, often have four or even six varieties in cultivation in the same holding.

In Karamoja and south Kigezi in Uganda sorghum achieves

a striking dominance, estimated amongst the Jie as 95 per cent of cultivation (Gulliver, 1954), but mostly occupying a half to two-thirds the farmed area, with finger millet (*Eleusine coracana*) as the chief minor crop and only very little cash cropping (McMaster, 1962, pp. 55–9). In southern Uganda the banana or plantain is remarkably dominant as the staple food crop, occupying 45–60 per cent of the food crop area in many counties (McMaster, 1962, p. 41). There are four main groups of varieties grown for beer, cooking and 'dessert', with sweet potato the chief minor food crop and coffee the chief cash crop. In McMaster's view the introduction and spread of coffee has strengthened the position of the banana as the chief staple because the two crops fit so effectively together into a system of cultivation. The insistence on banana cultivation, which is an important element in local tradition, has resulted in the spread of the crop into areas only marginally suitable, where other crops such as maize might have been preferred (McMaster, 1962, p. 104; see below, pp. 179–81). Maize dominance has reached an extraordinary peak in Central Mexico, where the map of maize distribution has reflected the map of population distribution, but where climatic conditions are only marginally suitable and where yields are less than the average for Mexico as a whole (James, 1942 and 1959, p. 651). Minor crops include beans, chile and alfalfa with linseed, groundnuts, sugar cane and maguey (the source of pulque, the popular alcoholic drink) as cash crops. Many communities have developed supplemental market gardening to supply nearby cities with vegetable crops or have very localized special crops such as onions or vanilla.

b. Two or more staples dominant

The most abundantly represented systems of agricultural production in the Third World are those with two or more staples and a commercial crop or crops. The advantages for labour deployment, soil management and husbandry discussed elsewhere (pp. 176–92) are considerable, and such systems also provide economic flexibility with a hedge against both economic and ecological risks. It is possible, however, that they may be slow to change as their variety in cropping and complexity in management provide difficulties for the introduction of new crops or the increased production of one particular crop without abandoning or making very fundamental changes in the existing system. It is this which in part explains some of the resistance to change experienced in certain areas. Often this is a resistance to fairly small changes when what is needed to effect progress is a wholesale advance with a promise of profit big enough to offset the effort and risks involved. Small-scale changes can be effected most easily where crop substitution is possible in the system as in the substitution of the commercial groundnut for the subsistence earth-pea (*voandzeia*) in West Africa or of maize for sorghum in many tropical locations.

The co-dominance of staples may occur mainly in order to provide an effective mixed cropping combination such as sorghum and pennisetum millet, as in tropical Africa and peninsular India, or a suitable crop rotation, which in many cases, more especially in tropical Africa, is a rotation of mixtures (see p. 232), or seasonal crop alternations, as, for example, in northern India, where cool season wheat, sometimes together with barley or sugar cane, alternates with warm, rainy season *bajra* (*Pennisetum typhoideum*) and *jowar* (*Sorghum vulgare*). In the last case the development of irrigation has not only made possible all-year-round cultivation but has encouraged the cultivation of commercial crops during the irrigated dry season, alternating with more tolerant coarse grains in part used as animal feeding stuffs during the rains (for sample detailed calendars see Blaikie, 1971 (II), pp. 17–18).

In tropical Africa there is a broad contrast between the root crop culture systems, mostly combining yams, cassava and maize with oil palms or vegetable crops for the towns as commercial crops in addition to sales of surplus staple production, and the grain producing systems of the areas with marked dry seasons, usually involving sorghum, pennisetum millet, finger millet, rice or *fonio* (*Digitaria* spp.) often combined with cassava, beans and groundnuts and with additional commercial crops such as cotton or tobacco. In peninsular India *jowar* and *bajra* are often co-dominant, with groundnuts and cotton as commercial crops. In Borneo the Dyaks of West Kalimantan rotate mixtures of rice, maize and cucumbers with cassava, beans and sugar cane, and develop small plots of rubber. Sometimes staple dominance is a temporary stage in the development of commercial cropping, especially tree cropping, systems. A peasant farmer in Ghana may intercrop plantains and cocoyams with young cocoa and for a time sustain a co-dominant staple system of plantains and cocoyam with cocoa as a minor crop. Eventually the intercropped cocoa matures and the food crops disappear so that the system becomes one of cocoa dominance. The same principle occurs in the *taungya* systems of timber production developed in Burma and India and later in tropical Africa, whereby temporary peasant farmers intercrop timber trees such as teak with food crops until the trees mature and food cropping is abandoned. The principle has also been applied for the expansion of coffee cultivation in Brazil, where the *colonos* or tenants develop new plantations by contracting to clear forest and plant coffee, and are permitted to interplant food crops such as maize, rice and beans for four to six years. Peasant organized staple food crop production marks the pioneer fringe of many expanding perennial crop regions, becoming more and more subsistent in character in periods of rapid advance with long lines of communication, often of poor quality, back to the major urban markets. Systems with several staple and commercial crops normally involve a variety of land-uses and of agricultural methods. These are partly to suit the different crops, partly because the crops may have different environmental preferences, more especially involving differences in soil and drainage, and partly because the cultivator may have to make preference choices for the location of his various crops in terms of soil, slope and accessibility. He may, for example, decide to cultivate preferred staples and commercial crops in near locations at high levels of intensity and stand-by or reserve crops in distant locations at low levels of intensity (see below, pp. 236–9). He may need to hire labour and may find that the most profitable proceeding is to use such hired labour for the cultivation of commercial crops, but, if abundant land is available, at low levels of intensity. An enormous variety of labour and land-use combination procedures exists, reflecting local marketing and environmental conditions and local traditions.

c. and d. Livestock systems

Systems in which livestock have provided an important commercial element in addition to subsistence production of either crops or livestock or both are gaining importance just as the growth of a class of higher wage and salary earners able to purchase meat has brought some degree of prosperity to many less developed countries. In the steppe lands, semi-desert and tropical dry savannas of the Third World, where nomadic pastoralism has long been the mainstay of small numbers of widely scattered people, traditional systems of livestock dependence, usually involving some exchange of animal products for staple grains, have acquired a modern commercial element with the sale more especially of cattle, sometimes of sheep and goats, to towns often at a considerable distance from the pasture lands. Commercialized systems of traditional origin of this kind occur chiefly in the drier portions of tropical Africa, North Africa and the Middle East, where the chief stimulants to change have been urban demand, the growing wealth of mineral, especially oil, producing countries and the building of transport networks. Their greatest development has been amongst Fulani pastoralists in West Africa who have established a considerable meat trade with the towns of perennial export crop areas in the south. They have established their own villages near to the main markets together with local fattening pastures, although most livestock move directly to market by road or rail. Some African pastoralists are reluctant to sell livestock as they fear any reduction in herd numbers, which represent a herdsman's capital. Some governments, such as that of Kenya, have offered incentives in the form of stable prices and field abattoirs, whilst the Uganda government even introduced compulsory sales quotas to prevent overstocking (Mittendorf and Wilson, 1961, pp. 17–18). In South and Southeast Asia very little meat is eaten, but livestock have a major role as draught animals, scavengers and producers of milk and eggs. There livestock are mostly a minor component of agricultural systems, subsisting in small numbers attached to each village on grazing areas unsuitable for cultivation and often provided with additional fodder. This is virtually unknown in tropical Africa, where very few livestock are owned by cultivators, with a few remarkable exceptions such as the Sérère of Senegal, who rotate pasture with groundnut and pennisetum millet fields. In Southeast Asia the water buffalo has a most important role in agriculture, providing the draught power required to plough the wet paddy lands. By contrast in Africa and in tropical Latin America the plough is little used. In Latin America pure livestock rearing is almost entirely a commercial operation on large ranches, but livestock rearing on farms and smallholdings in combination with crops has played an important role in pioneering agricultural settlement. An important instance is in southern Brazil, where German colonists created maize and pig farms. In the north-east cattle ranching has been combined with cotton cultivation where tenant cotton farmers on short-term contract have cleared the *sertão*, the interior grasslands, and *caatinga* scrub-forests, making them suitable eventually for grazing. In the Argentine pampas wheat and cattle raising have been combined in a related manner by renting portions of the cattle *estancias* to tenant farmers for 5 to 10 years and allowing them to grow wheat providing they also grow alfalfa for cattle feed (James, 1959, p. 340).

2. *Mainly subsistence*

Pure subsistence, that is the production of all that is needed for survival on one's own holding, is extremely rare and may hardly be found outside a few extremely remote areas of Southeast Asia. Usually some limited occupational specialization and exchange exists even if locally confined and effected mainly by barter. Contact with the outside world inevitably results in a widening of exchange, often with further specialization and in effect a narrowing of the range of production. The concept of a subsistence agriculture is generally misleading. Rather there are a few nearly subsistent economies, in some of which, as one might expect, there is a great variety of cropping enterprise in order to satisfy local needs, whilst in others needs are either more limited or are satisfied by few crops, but supported by gathering and hunting. Cultivation in such conditions is normally extensive, shifting through an area of fallows which may range from 6 to over 20 years. Fürer-Heimendorf (1952) has described the Jen-Kurumba or Naikr, a forest community living on the borders of Mysore, who plant *ragi* (*Eleusine coracana*) and a few vegetables, use primitive slash-and-burn methods, shift their settlements from time to time and are dependent on gathered produce. Freeman (1955) has described the Iban of Borneo with their varieties of upland rice. In the central Highlands of New Guinea at altitudes of 4000–9000 feet above sea level the degree of isolation is extreme and contact with outside communities has been very recent. Here some 900 000 people of Neolithic culture have come into contact with the Western world only in the present century, in some instances not until the late 1950s. These people have an extraordinary dependence on root crops, more especially on the sweet potato, either as the principal or as the second crop, which in many localities provides over 80 per cent of the food intake. In the Chimbu area alone there are at least 40 varieties. The chief alternative principal crop is *taro*, the Colocasia cocoyam, and minor crops include sugar cane and bananas, yams, especially at the low altitudes, manioc (cassava), legumes, green vegetables and a few tree fruits, particularly the nut pandanus. Pigs are kept and frequently used as currency. Fallows of fifteen years or more are not uncommon, but cultivation may continue for several years before fallowing and the fallows are often planted with casuarinas or *pitpit* grasses (*Saccharam spontanaeum* and *Phragmites karka*) to restore soil fertility. In addition to simple shifting cultivation methods, however, Brookfield has recognized five other classes of more intensive cultivation and has suggested a development of techniques in response to growing population pressure and the greater fertility of high-altitude soils. The new techniques made possible higher densities of population without improving the level of subsistence (Brown and Brookfield, 1959; Brookfield, 1961, 1962 and 1964). Already, however, the system is changing with outside contact. Coffee is now a major cash crop.

In the South American Andes one can recognize a maize dominant subsistence agriculture supported by minor crops such as *yucca* and black beans. Sometimes the 'rotation' is ended with plantains and bananas. The *conuco* farmer of Venezuela is a migrant, using simple tools such as the machete and digging stick, often a pioneer on new lands, but with extremely low levels of productivity and frequently regarded as a destroyer of forest and a cause of soil erosion. Many such farmers are *ocupantes* or squatters on large estates. Some plant small areas of cash crops, such as coffee, although commercial cropping in combination with subsistence cropping is done mainly by more settled peasant farmers, usually share-croppers and renters

(Watters, 1966 and 1967). Many others are part-time farmers employing the *conuco* in order to supplement their income from wage labour. In Venezuela most cultivators appear to be partial subsistence farmers on *conucos* and partial wage labourers or commercial farmers on more intensive farms. Single crop dominant subsistence farming often seems to be associated with mountains, but there is no general rule. Brookfield (1964), for example, contrasted the multiple cropping of the highlands of northern Luzon with the cropping systems of the Central Highlands of New Guinea and concluded that the New Guinea highlands were more narrowly restricted than they would have been with a more varied cropping pattern and food storage.

For the most part subsistence cultivators employ a great variety of crops and usually have two or three co-dominant staples. In West Africa, for example, the Boki, Ekoi and Anyang depend on the shifting cultivation of plantains, cocoyams and maize which combine usefully in a crop association which makes possible a high density of production, thus minimizing clearance and spreading work loads. In the south-west the central Gouro and some of the Kru have upland rice as a single dominant, but other Gouro and Kru together with Gagou and Dan plant both yams and plantains. Subsistence farmers dependent on systems with little or no fallow and planting pennisetum and digitaria millets, sorghum and, in the east, finger millet (*Eleusine coracana*) still exist in many of the upland regions of the West African 'middle belt'. Today the ancient subsistence systems are giving way to the influence of commercialism associated with resettlement in the lowlands and the emigration of many young people to the towns (Morgan and Pugh, 1969, pp. 103–4). In rainforest areas cocoyams and plantains or maize and rootcrops, especially cassava or manioc, are common co-dominants. Such crop combinations for subsistence occur in West Africa, the Zaire basin, Angola and Brazil. In Angola maize is often combined with manioc, sorghum and bulrush millet by the farmer-herders of the Huila Plateau, who raise cattle chiefly for their milk supply (Urquhart, 1963). In Eastern India, Bangladesh and Assam subsistence agriculture is frequently associated with long fallows and clearings or *jhums* worked for periods of up to five years, within which many peoples sow rice, pennisetum millets and job's tears (*Coix lachryma*) broadcast, often in association with minor crops such as black dal, peppers and cucumbers. In the hill lands of Dominica the *conuco* slash-and-burn farm or *parcela* is devoted chiefly, often almost entirely, to subsistence crops, including maize, manioc and plantains as co-dominants, together with sweet potato, pigeon pea, squash, broad beans and some tobacco for local use. Hill land is abundant and extensive methods of cultivation prevail, including leaving the stumps and roots in the ground (Antonini, 1971).

The subsistence raising of livestock is rare. Even in the past there was usually some exchange for grain or vegetable foods. Perhaps some of the Masai of East Africa, living traditionally off the milk and blood of their cattle, and judging a man's social status by the size of his herd, have come nearest to the ideal, although even amongst the Masai the trading of livestock products for grain with, for example, the Kikuyu has not been uncommon, and amongst some Masai the women have traditionally practised a limited form of cultivation.

Subsistence cultivation is essentially a cultivation of the fringe lands or more isolated locations of the world. It is in consequence often associated with areas of low density of population, where markets and transport have been slow to develop and with few nucleations of settlement. In such conditions inputs are normally low, unless local overcrowding through warfare or environmental limitations, occurs, and agricultural techniques are frequently at their simplest level. It could be argued, and it is certainly implied in Brookfield's thesis (1975), that subsistence cultivation may be seen in consequence as the product of development, a way of life forced outwards to the perimeter of human contact and in many cases this view is well supported by the evidence, particularly in Latin America, where many farmers choose subsistence because they are deprived of commercial opportunity or are unwilling to accept the conditions imposed on tenants or labourers of the huge estates. The spectacle of low density, under-used *hacienda* lands, alongside extremely over-crowded areas occupied by peasant farmers, many of whom feel forced to migrate either to *conuco* cultivation on the pioneer fringe or to under-employment in the towns, is evidence enough. In tropical Africa and in India there is decline or even the virtual destruction of ancient commercial systems, of former handicrafts and of social systems which depended on locally produced food surpluses. These have been replaced by port oriented systems associated with a new commercialism. In Northern Nigeria the British conquest resulted in the abandonment of towns and villages, occupied mainly by slaves who formerly worked on estates, and who returned home to a mainly subsistent and more isolated existence until new roads and markets were created. There is also evidence, however, of a traditional subsistence cultivation in numerous locations in the Third World, which was not created by modern development nor pushed into its present location by the spread of modernization. Such subsistence systems occurred in areas which have since become fringe lands. Their status was, as it were, confirmed rather than produced by the new system of world exchange.

3. *Mainly commercial*

a. Family and hired labour

The commercialization of smallholder production in the Third World has been largely the result of the development of overseas trade and this has been true even of the commercial production of food crops for the local market, since much of the demand has been generated by a prosperity created by the developing export economy. The rapidity of development has often been astonishing, resulting, more especially in the great period of export economy expansion from about 1890 to 1929, in the planting of many millions of trees to provide produce for sale chiefly to Western Europe and North America, and the creation of considerable transport networks, market systems and urban hierarchies. Throughout the Third World such developments have been oriented towards the coasts, at first towards a number of small ports and later concentrated into fewer ports as the production and exchange systems expanded. These in turn involved greater dependence on transport networks as a developing port technology created a few special systems of handling goods and as it became more and more desirable to control trade through a few ports with adequate customs supervision (see below, pp. 208–9). Much of this development was a response to a considerable market incentive, which in most Third World countries could come only from overseas, where the expansion of manufacturing industry created a demand both for special tropical products and for foodstuffs for a largely urban population. In some regions the effect has been dramatic, notably in Brazil with its notorious boom crops and 'hollow-frontiers' of commercial crop expansion; in the cocoa regions of West Africa, where less than half the income, in some instances less than a fifth, comes from local food crops; in the island of Mauritius, where 80 per cent of the arable land is in sugar cane; and in Senegal, where on most holdings two-thirds or more of the area is in groundnuts. Most highly commercialized smallholder systems, especially those concentrating on export crops, are near major ports, or, if distant, depend, like groundnut production in Northern Nigeria, on especially advantageous local freight rates.

Most commercial smallholdings in the Third World are family farms of 2 to 20 hectares concentrating on one major commercial crop, which is the chief cash earner of the district, with an elaborate system of marketing and often associated with government supervision of quality grading and the levy of export duties. A minor commercial crop may also be grown in order to avoid complete dependence on a single crop for cash earnings, and a number of mainly subsistence food crops are grown, sometimes interplanted in the same field as the major commercial crop. The commercial crop may not occupy the greater part of the farmed area; nevertheless it will dominate the farm economy and its system of operations. The family provides most of the labour, the chief input in most systems which operate at a low technological level and are normally not affected by land shortages. Commercial crop production on smallholdings does exist in overcrowded areas and may even have encouraged overcrowding, but generally the more overcrowded the farmlands the more important the subsistence food crops and the more 'mixed' are the agricultural systems. Important exceptions like groundnut cultivation in Cayor, Senegal, or sugar cane growing in Mauritius have been related to systems of market stabilization which have offered steady returns and have often been linked, as in the former French imperial system of tropical crop production, with a supply of cheap imported foodstuffs (see below, pp. 185–6). Few do not hire labour, which is essential to relieve bottlenecks in the labour supply at peak periods of work load and for the expansion of production at times of economic boom. Smallholders frequently use both their family labour and their land inefficiently, but may have little or no alternative use for such labour and are therefore prepared to accept low returns for their effort. In many cases they devise highly intensive systems of production, employing a skilfully chosen combination of crops both for market and for a satisfactory use of soils. They produce most of the commercial crops of the Third World, including the export crops, despite a dominance by plantation systems in the past, and their share of commercial cropping is rising. Like plantation cultivators they have had most success with perennial crops, particularly with tree crops such as cocoa in Ghana, Nigeria and Trinidad, coffee in the Ivory Coast, Kenya, Uganda, Colombia, Angola and Madagascar, rubber in Liberia, Nigeria, Malaysia and Sri Lanka, oil palm in Nigeria and Malaysia, coconut in the Philippines, Sri Lanka and Samoa, and various fruits, especially oranges. Tea production is still dominated by the plantations. Smallholders also grow perennial field crops such as sugar cane, pineapple and sisal, a category into which bananas may also fit to some extent, and annual field crops, including groundnuts, cotton, tobacco, pyrethrum and an enormous variety of foodcrops, mainly for local markets. Contrary to the impression given in some textbooks, a large part of the coffee production of Brazil comes from *sitios* or smallholdings, growing coffee in pure stands, but also growing food crops and raising cattle. Casual labour is hired for the harvest period and extensive methods are used, keeping costs to a minimum (Ruthenberg, 1971, pp. 224–7). *Sitios* are especially characteristic of the most recent areas of production in Parana. Perennials have the advantage for smallholders of ease of management and require few material inputs other than labour. Their disadvantage is the need for enough capital to acquire seedlings and to survive until the trees come into full bearing, which may take seven to fifteen years. This can be partially offset in the early stages by intercropping food plants which may also act as shade for the seedlings. Considerable research has been concentrated on the development of quick-maturing trees in order to solve a major smallholder problem. The Rubber Research Institute of Malaysia, for example, has been trying to develop a tree which can begin to yield in 3 years instead of 5 and at 2–3 times the output per hectare. The economic risks are high since the returns are so long term, but labour inputs in the later stages are very low and many smallholders prefer to allow an old 'plantation' to run down and in a period of boom to plant a new holding many miles away on the frontier of the expanding crop area. Many smallholdings produce two or more cash co-dominants in systems in which the proportionate importance of each major commercial crop is either nearly equal or allowed to vary according to the estimated future state of the market. Some interesting combinations are coffee and cocoa in the south-eastern Ivory Coast, where the two crops are planted on different soils, bananas and coffee in Bukoba and on Mount Kilimanjaro in Tanzania, cocoa and kola (*Cola acuminata*), the latter producing a nut chewed for its drug and supplying the local market, in south-western Nigeria, and rubber and pineapples in Malaysia. Many commercial smallholdings are too small to support a family and are consequently operated part time. In such cases the smallholder seeks work elsewhere, in the town, or on large farms employing labour. Such is commonly the situation in overcrowded lands in much of India, Bangladesh and Pakistan, where many holdings are so small that intensification is no longer worth while and most effort is put into increasing the income from other sources. In Mexico the *rancho* or smallholding is often only 3 hectares or less and in many areas their owners must work for others to support their families, often in debt bondage to the owner of a *hacienda* or large estate.

b. Management and hired labour

Farms which are not family firms but managed systems with all labour hired, whether permanent or casual, are normally very large in the Third World and some are amongst the largest agricultural holdings in the world, as for example in Liberia where the Firestone Rubber Company leased approximately 400 000 hectares in 1926 and eventually developed a rubber plantation at Harbel of over 56 000 hectares. Usually centrally managed farms or plantations have only a small proportion of their area planted to a major commercial crop, even though that small proportion may still be several thousand hectares, and the rest is allowed to be idle or to support food-crops or woodlots or to be occupied by squatters. Most plantations have been created by Western capitalism, but some are the product of Third World initiative, including that of Government agencies sometimes based on a socialist model such as the many Soviet inspired state farms. Essentially they are large units employing a high degree of specialization and a sharp division of labour, even where, as occasionally happens, they may produce more than one crop. A characteristic feature is the intensive use of labour for cultivation processes difficult as yet to mechanize (Courtenay, 1971, p. 123). Most specialize in just one crop and have experienced greatest success with perennials, particularly tree crops, or with shorter-term crops such as sugar cane and the banana. They have been located mainly in areas of low population density where large areas of cheap or even nominally priced land were available for exploitation. In consequence, unless they were close to an area of high concentration of population, like the coffee plantations on the periphery of the Kikuyu highlands of Kenya, they were forced to import labour, often from considerable distances, and develop their own road, services, supply and marketing systems. Often they have provided a legacy of social and political difficulties by introducing large numbers of people of alien culture such as Tamils from India into Sri Lanka for coffee planting and tea picking and into Malaya to tap rubber. Plantations are not exclusive to colonies or former colonies and although the number privately owned has decreased in recent years, the number of state owned, large, centrally managed units has increased considerably.

Whether privately or state operated, labour costs on plantations compared with those of local smallholdings tend to be high and so are their overheads, especially in housing, power, machinery, roads and high quality plantstock. Their greatest success has undoubtedly been with crops where considerable preparation is required, where high standards of quality are paid a premium and where central control combined with labour division achieve substantial economies. Plantations are still amongst the most important producers of sugar in the West Indies, bananas in Central America and Cameroun, rubber in Liberia and Malaysia, tea in Assam, Sri Lanka and East

Africa, palm oil in Zaire and Malaysia, and sisal in East Africa. Many are owned and operated by multi-national corporate firms who reduce their political risks, which today are amongst their largest, by spreading their operations amongst a number of countries, and their economic risks by diversifying their interests. Many of the firms have a greater financial turnover than that of the total agriculture of the countries within which they operate (Beckford, 1969). Many plantations have been nationalized and whilst many have been subdivided into smaller properties, others have been preserved and even expanded. The centrally managed farm idea has now been developed for other forms of crop production such as food farming in countries where high import bills for food are regarded as the result of failure of peasant production methods. Large state food farms have been developed in Tanzania and in Ghana, sometimes in the place of former Colonial Development Corporation schemes, chiefly for groundnut production (La-Anyane, 1974). In Ghana they have largely failed and in Tanzania much of the former groundnut area is now in small tenant farms. Large units have frequently been regarded as necessary for the economic use of machines, which are essential to expand food production where labour is concentrated mainly in other economic activities (see pp. 164–6).

Large scale centrally managed livestock systems or ranches have occurred mainly in areas of European settlement, particularly in temperate and sub-tropical grasslands, mostly in developed countries. Important Third World ranching developments occur in the Argentine and Uruguayan pampas, the *llanos* of Venezuela, the *sertão* of Brazil, Patagonia and the Argentine Chaco, in a few instances in tropical Africa, more especially in West Africa, where Government ranches have been created to meet a growing demand for meat, but hardly any in South or Southeast Asia, the Middle East or North Africa, where traditional livestock raising systems still prevail. Most ranching land is of low productivity. Little attempt is made to intensify and most systems tend to keep costs to a minimum and rely on unimproved natural grazing. Normally 2–10 hectares per beast are required, although in the Argentine pampas less than one hectare is required on *estancias* combining wheat with lucerne as a planted fodder. Wheat became a major product of

Argentina partly as a by-product of ranching (Grigg, 1974, p. 249). In Argentina and Uruguay beef production is one of the most important economic activities and a major export industry, but elsewhere in the Third World the production of low-grade beef and mutton on ranch lands is intended chiefly for a home market of limited size and requiring cheap meat; hence low-grade meat production in West Africa on government ranches in 'derived savanna' near to major cities, but competing with the product of pastoral nomadism on more distant open grazing lands in the Sahel; and hence low-grade meat production in the Brazilian pioneer fringe and in the alternating drought and flood conditions of the Venezuelan *llanos*. Partially centrally-managed systems can occur where traditional methods cannot be easily improved but where Government agencies can manage fattening or holding grounds near to market where livestock are fattened before slaughter. Such fattening areas have been developed on a large scale in Rhodesia by the Cold Storage Commission and on a smaller scale elsewhere, notably near Fort Lamy in the Chad Republic, on the Niger flood plains in Mali and in the Republic of Niger.

c. Centrally-managed rent or sharecropping smallholder systems

Elsewhere central management has been combined with smallholding either to achieve a very cheaply run estate in which overheads and labour costs to the owners are reduced, or a combination in which a central system of support and advice ensures a greater degree of success for smallholders. In many cases mixed systems have emerged such as plantations employing wage labour combined with family smallholdings on owner-occupied, rented or leased land jointly supplying a central processing plant and trading system. Such mixed systems have even been advocated in the form of 'nucleus estates' in which a plantation is encouraged by government to install processing capacity beyond its needs and therefore able to absorb the output of smallholders on its periphery. The notion is to combine greater scale economy in processing with the supply of superior planting material and the most recent technical information to smallholders (Phillips, 1965). The difficulties in developing such systems would in part be political, for investors might feel that government would be more concerned with the interests of smallholders than with theirs, and partly socio-economic in that such arrangements in the past have often been achieved in boom times when plantations sought additional output to process, whereas in times of slump they concentrated entirely on their own production, that is they tended to use the smallholder system as a means of introducing a degree of flexibility into their operations.

The cheaply-run, low-input system estate with centralized management and smallholder tenants or share-croppers is especially characteristic of Latin America. Share-cropping has been especially important in Brazil, for example, as a means of preparing coffee or cocoa estates for production. The share-cropper combines maize, cotton, beans or rice with the young permanent crop, thus spreading weather and price risks (Schuh, 1970, pp. 132–3) and providing a very cheap system of preparation. As export crop perennials play a less important role in the economy and as the coffee frontier has moved southwards where the use of animal and mechanical power is more important, so share-cropping has declined. In São Paulo part of the coffee crop was tended by *colonos* on the *parceria* system, usually recent immigrants hired annually for periods of up to five years to look after the coffee trees of a *fazenda* and allowed to grow maize and beans on a separate plot. After five years the coffee *fazenda* is operated either as a centrally-managed system with wage labour or as a collection of tenant farms. Often the owners are absent in the city. More and more *fazendas* have been subdivided into farms of less than 60 hectares and sold off to tenants or other small owner-occupier farmers who have developed more diversified farming systems, often with the emphasis on food crops, in the wake of the advancing coffee zone. Encouragement to do this came partly from difficulties of re-establishing coffee on former coffee land and partly from the huge influx of immigrants who moved into farming as a temporary stage in their more general movement towards Brazil's major cities. Their establishment on part of a *fazenda* was regarded by some owners as an important reservoir of labour to relieve periods of peak demand (James, 1959, p. 490, and below, p. 216). In north-eastern Brazil the sugar *fazendas* mostly have absentee owners but with more permanent tenants than the coffee *fazendas* of São Paulo. These also provide a wage labour force for the cane processing mill. In Cuba the sugar 'plantations' were mainly worked by tenant farmers who were tied to the estates by their dependence on the sugar mills or *centralos*. Even small independent farmers who grew cane were tied to the nearest sugar mill. The tenants now work on Government estates following the confiscation of very large estates in the early 1960s. Throughout Latin America most *haciendas* and *fazendas* were and still are estates operated by share-croppers or tenant farmers with some hired labour, rather than centrally operated plantations. The *hacienda* lacked capital and monopolized land in order to monopolize labour whose organization was the *hacienda's* chief innovation. The *hacienda* has been essentially a conservative organization preserving traditional peasant methods of production, even in some cases subsistence production and shifting techniques, extensive rather than intensive in its operations and profiting by low costs rather than high output. In Mexico the *hacienda* proved 'a self-limiting economic system, incapable of further expansion' (Wolf, 1956, pp. 56–9) and elsewhere its decline has frequently been hastened by government realization that the *hacienda* system is in large part a bar to greater agricultural productivity.

Elsewhere in the Third World estate systems of the kinds found in Latin America are much less common, although share-cropping arrangements do occur in West Africa, where some larger holdings are cleared and developed for cocoa, coffee or rubber by short-term farmers who are permitted to plant subsistence food crops on the holding, and in West Malaysia where Chinese 'squatters' plant pineapples and clear weeds for a monthly wage and a share of the profits from the sale of the crop (Wee, 1970). Many large estate systems are government owned and operated and represent attempts to create systems which are efficiently managed, innovative and combine economies of scale with traditional organization. An outstandingly successful scheme of this kind is that of the Gezira in the Sudan. Together with the Manaqil extension it consists of 755 000 hectares of irrigated clay plain divided into tenant farms mainly of just under 17 hectares, a quarter in high quality cotton, and a quarter in sorghum and a fodder-crop, usually *lubia* (Lablab niger), together with a half uncultivated.

Even units of this size are large for family labour and need the assistance of hired labour at peak periods of work load. More recent holdings are smaller, even as small as 6·3 hectares, and more intensively worked, and several of the older larger tenancies have been reduced in size. The management and the tenant farmers are in a partnership in which cultivation routines and types of crop are controlled by the management, but in which tenant farmers may expect incomes well above the national agricultural average (Barbour, 1961, pp. 200–7 and below, pp. 246–50). A similar scheme of partnership between tenant farmers and Government in Mali in order to produce rice and cotton, in part by the use of mechanical implements, has had less success and proved extremely expensive (De Wilde, 1967, I, pp. 72–4, II, pp. 245–300; Morgan and Pugh, 1969, pp. 645–56). The Sudan also has one of the few examples of a successful centrally organized tenant scheme in which mechanized methods play a major role. The Mechanized Crop Production Scheme is on rainlands, and operated by wealthy tenants who supply their own tractor and implements, employ farm managers, and practise a form of long-term fallowing which resembles shifting cultivation (Davies, 1964).

THIRD WORLD CROP AND LIVESTOCK PRODUCTION FOR INTERNAL MARKETS

Many authors distinguish between food crops and cash crops, the latter often being regarded virtually as a synonym for export crops. Although in many Third World countries subsistence food production may be the largest element in the production system, nevertheless such a distinction is not particularly useful and may be invalid. Frequently the most important export crop is a food crop, or the chief cash crops are those sold for local consumption and not for export. Some crops overlap the categories chosen, inevitably when there is more than one 'market' for a particular crop and in some cases more than one use to which a crop may be put. The distinction drawn here is between crops for the internal market, including subsistence crops, and crops for export. This distinction is important because of the very different sizes and nature of the markets involved and the different policies and organization of production and marketing which have occurred.

At the turn of the century the main emphasis in attempts to develop Third World agriculture was on export crops, partly in order to earn foreign exchange, partly because of the appreciation of the widening scope of the world market for tropical products and partly for taxation and government revenues. Today there is much more interest and even insistence on expanding crop production, more especially of staple foods, for the internal market. Food productivity cannot be separated from the problem of general productivity. The argument that a decline in world trade could even be an advantage to many Third World countries because the peasant farmers would concentrate more on food crops, whilst not entirely without substance is nevertheless of little value when one considers the tendency towards very low levels of production of many 'subsistence' farmers and the increased market dependence of the rapidly growing Third World urban populations whose condition is likely to be seriously worsened by a decline in trade. Moreover, the reasons for so-called failure to satisfy home food demands vary enormously. Some countries are subject to high variability of moisture supply or of disease or pest incidence so that their production is subject to irregular shortfalls. Others, more especially in southern Asia, have huge populations pressing hard on a fragile agricultural resource base, whilst yet others have created dependence on overseas food supplies by their success either in directing their agricultural production towards export crops or their success in developing other sectors of the economy with a rise in standards of living which was satisfied more easily by importing consumer goods than by producing them locally. Moreover, it has almost always been easier to increase production in response to an expanding market in the more developed countries well able to pay for their imports, than to respond to the needs of poor people in a weakly developed market situation where growth involved considerable new expertise and the building of a new physical and technical infrastructure.

As already indicated increase in agricultural production in the Third World has so far been met more by increases in the farmed area than by increased intensity of effort. As increased intensity normally involves higher costs, which few Third World markets are capable of sustaining, this is hardly surprising. The Third World as a whole has a rough parity in available land with the world as a whole, for in 1973 it had 45 per cent of the world's arable land supporting 48 per cent of the world's population. In many Third World countries, however, more especially in overcrowded Southeast Asia, and even including some African and American countries with generally low densities of population, the best lands have been taken and further expansion will mean poorer returns for comparable effort. Much of this expansion has been subsistence oriented and has resulted in a scatter of discontinuous islands of agricultural settlement (Grigg, 1970, pp. 44–5) which will be costly to link into a modern commercial system should agricultural innovation and access to markets be desired. There are few extensive tracts of arable land in the largely tropical Third World and most of them are in the very wet or riverain lowlands of southern Asia, more especially in India, Pakistan, Bangladesh and Sri Lanka with an estimated 13 per cent of the world's arable land supporting 18 per cent of world population.

The leading field crops of the Third World for satisfying internal needs are in approximate order of crude tonnage produced: rice, cassava or manioc, wheat, maize, plantains or bananas (including commercial production for export), sorghum, potatoes, yams, pennisetum millet, groundnuts, cotton and cocoyams. All of these with the exception of cotton and groundnuts are starchy food staples. In addition there is an enormous number of other field crops including staples which locally may be of considerable importance such as eleusine or finger millet, and sweet potatoes; fibres such as hemp, ramie and kenaf; oilseeds such as sesame, sunflower, soya beans and castor; a variety of legumes and leaf plants; drug crops, fodder crops and even poisons. The agricultural plant stock of the Third World is in practice much more varied than that of the more developed world and its potential for improvement is enormous. The variety is, however, partly the result of lack of development in certain countries where traditional crops have persisted partly by cultural isolation and lack of incentives for change. Despite the world wide dissemination of crop plants, particularly in the late nineteenth and early twentieth centuries when an enormous interchange took place between Latin America, Africa and Southwest and Southeast Asia, there is evidence of local decreases in the variety of plant stocks, as many local crop plants were replaced by exotics or went out of production in an age of increased specialization. The regionalization of Third World cropping patterns is a process which seems likely to continue and to increase as development takes place.

Rice, the leading crop of the Third World, which in 1975 produced nearly 192 million tonnes on 97 million hectares (Fig. 14), is one of very few crops which in certain localities are subject to intensive methods of production, mostly involving water control, and often involving high labour inputs and the use of fertilizers. Even so Third World yields at 1980 kg/ha in 1975 were well below the world average of 2441 kg/ha, and the rest of the world produced over three-quarters as much rice again (including production in China and Japan). Although rice may be grown as a rainland crop with low yields, as it frequently is in Africa and the uplands of Southeast Asia, it possesses the property unique amongst cereals of being able to germinate and thrive in water. In floodland conditions a much better control of environment is achieved and the maximum advantages may be gained from the available heat and light. Rice is of tropical origin and has achieved its most remarkable development in Southeast Asia, but within the tropics a high incidence of disease and low levels of fertility either from soil conditions or from limited use of fertilizers have resulted in lower yields than those achieved in warm temperate Australia, Spain, Italy and California, which enjoy the additional advantage of longer hours of daylight in the growing period. The rice map is determined essentially by culture and population. Undoubtedly Southeast Asian farmers have moved to the wet lands where the highest levels of productivity of *Oryza sativa*, the most successful species, have been achieved, but paddy farming in *sawah* requires a large labour force per unit area and therefore a high density of population in the area farmed. A large part of oriental rice is grown by transplanting from nursery beds, increasing yields and making possible two crops a year in some especially favoured areas. Such areas now have severe problems of agricultural development as population increase has

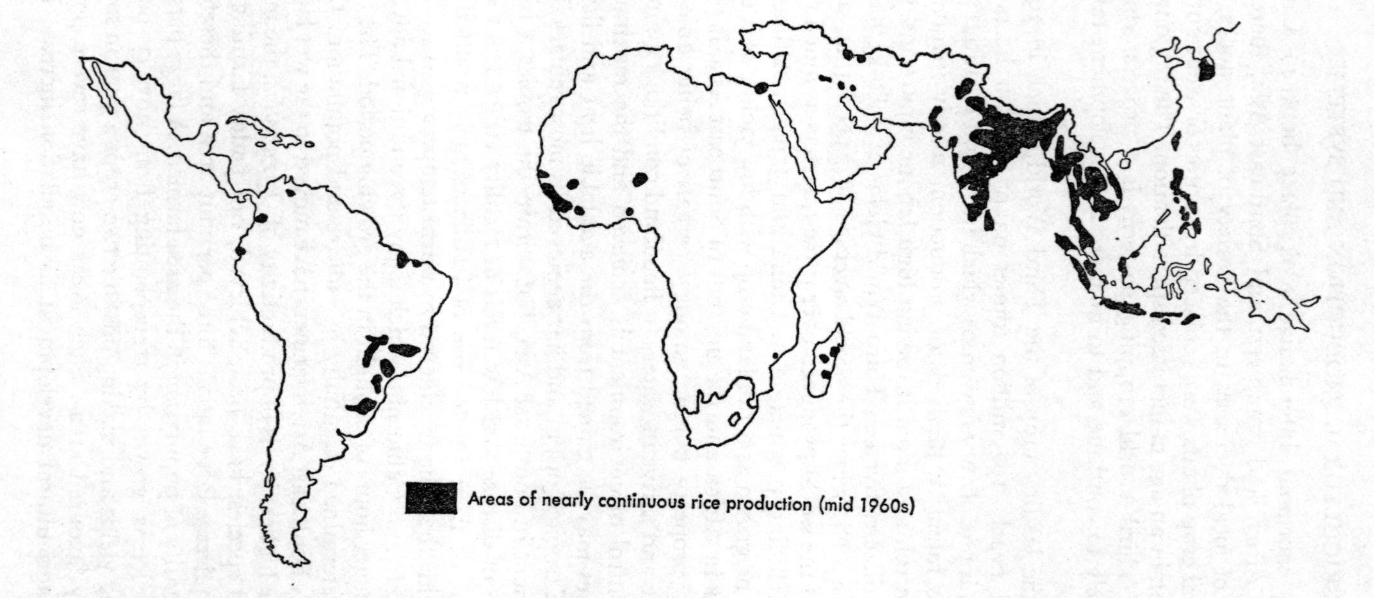

Fig. 14. Main areas of rice production in the Third World

pressed on the high levels of productivity achieved by special techniques, making the possibility of change to other forms of agricultural production extremely difficult. *O. sativa* has spread elsewhere and quite importantly to Brazil, Madagascar (where it was introduced by the Merina, sometimes referred to as the Hova, a people partly of Southeast Asian origin) and West Africa, where it has partly replaced the indigenous *O. glaberrima* which tends to shed easily.

Cassava, manioc or mandioca (*Manihot utilissima*), a native of Brazil, the source of the tapioca starch of commerce, is probably the most productive root crop in calorific terms, capable of yielding more than 80 tonnes of starchy root per hectare in conditions of intensive cultivation, using fertilizers. The crop is a major staple, particularly of subsistence farmers or commercial crop farmers producing most of their own food requirements on rainlands in tropical Africa and Latin America and to a lesser extent in Southeast Asia and Oceania. In many cases it is a standby crop to which little attention is given and it is mostly grown, even as a main crop, as a hardy staple which will yield well in a variety of soils and with little attention. Average yields therefore tend to be low, varying regionally between 7 tonnes/ha in Africa and 14 tonnes/ha in Latin America. Because of its importance as a major secondary crop it is widespread in distribution yet confined almost entirely to the tropics and largely absent from areas where floodland rice is the staple. It has been estimated that the annual production of cassava from tropical countries is more than half the total root crop production, supplying enough calories to satisfy the total energy requirements of some 200 million people and therefore the staple food of probably 400 millions (Coursey and Haynes, 1970) or nearly 22 per cent of the population of the Third World. Unfortunately despite its obvious importance the distribution data are unsatisfactory for mapping. Cassava does not require seed, being propagated from stem cuttings, nor storage since it can stay in the ground without rotting for periods of up to three years. Although its varieties take 8 to 15 months to mature it can spread its growth over several rainy seasons and thus grow in short rainy season areas where the immunity of the root to locust attack has often proved a useful property. Cassava flour can keep for long periods and has proved popular as a cheap food in African towns. It is, however, almost entirely lacking in proteins and vitamins and its use in weaning infants in tropical Africa has been claimed as the major cause of the infantile protein deficiency disease known as *kwashiorkor*.

Although less than 23 per cent (1975) of world wheat production comes from Third World countries nevertheless it is the third most important crop in tonnage. Its most important areas of production (Fig. 15) are in the warm temperate and subtropical margins of the Third World, more especially in northern India, Pakistan, the Middle East, northwest Africa, northwestern Mexico, Argentina, Uruguay and Chile. For an important food staple in the Third World wheat is therefore remarkably restricted in distribution and mainly by environmental factors, more especially by the combination of high rainfall and high temperatures which characterize much of the Third World and which favour the development and spread of wheat diseases, encourage lodging and leach nitrates from the soil. The lack of cold winters discourages the cultivation of high quality bread wheats. Nevertheless the popularity of wheat bread continues to increase rapidly throughout the Third World, so that imported grain and flour are major items of trade, usually North American hard wheats sometimes imported under aid arrangements. In some African countries baking has become the most rapidly growing manufacturing industry. Attempts to develop import substitution by home-grown wheat are almost universal and experimentation with new varieties and more especially with dwarf hybrids is virtually universal. The areas chiefly favoured are the more temperate tropical highlands and drier regions with favourable irrigation prospects.

Maize is as widespread in geographical distribution as wheat is restricted, but unfortunately there is only a slight indication of this in figure 16 since maize occurs chiefly as a major secondary crop or minor staple throughout the humid tropics and has a highly discontinuous distribution. It is currently making rapid advances into temperate countries as a remarkably high yielding fodder crop and a minor foodstuff in the form of sweet corn. In part the reason for this spread is the extraordinary diversity of maize varieties developed by the

American Indians, so that maturation ranges from seven weeks to almost a year, together with the modern development of a vast range of hybrids. Mostly, however, higher yields are achieved in areas with a growing period of at least twenty weeks. Within the broad spread of maize there are islands of major concentration of production, including humid tropical concentrations in southeastern Brazil, central America, Java, the Philippines, East Africa and southwestern Nigeria and Benin Republic, almost entirely for subsistence or the home market, in subtropical to warm concentrations in central Mexico, northern India and northeastern Rhodesia, mainly for home consumption, and in the Argentine pampas largely for export as a livestock feeding stuff. The reasons for each concentration are extremely varied and depend in large part on the relationships of maize to other food crops, patterns of diffusion and cultural traditions. In some areas it was an important substitute for sorghum and valued both as a bread grain and a source of beer or spirits; in others it became an object of almost religious veneration, an easily grown feeding stuff for slaves, cheap hired labour or livestock; in others again several varieties have proved valuable in combination with other crops for cultivation in floodland. Quick-growing maizes are important in subsistence cultivation throughout the tropical world as hunger breakers and in more equatorial latitudes make possible double cropping on rainlands.

Information on the cultivation of bananas as a local staple foodstuff, usually in a starchy form intended for cooking and commonly called the plantain or platano, is difficult to obtain. Most data refer to the sweet bananas of world commerce. Plantain varieties are largely restricted to the humid tropics, more especially to areas with rainy seasons of nine months or more, and are chiefly associated with the equatorial rainforests. Yields are extremely high, generally averaging 8 to 18 tonnes per hectare even in subsistence cultivation conditions. Planting is by suckers producing fruit in a year to a year and a half and with manuring a clump may continue to produce for five years or more. Cocoyams, the edible aroids (*Colocasia esculenta*, taro or eddoes and *Xanthosoma sagittifolia*, tannia) have a similar preference for equatorial rainforest conditions, and not only provide shade for young plants but prefer shade themselves, thus making a suitable intercrop with tall perennials and very frequently with plantains. The plantain-cocoyam association is important in many locations, more especially in equatorial Africa.

The grain sorghums (*Sorghum vulgare*), although of less importance in the Third World than rice, wheat or cassava, are nevertheless among its most characteristic crops, having a close association with the distribution of the humid tropics with a marked dry season, that is with the savanna lands, and being a universal staple of subsistence farmers. They are also difficult to map. The many thousands of varieties (best known are *dura*, *feterita*, kaffir corn, guinea corn and *kaoliang*) include bread grains, brewing grains (the chief source of locally brewed beer in Africa), livestock feeds and sweet grains for chewing. Some sorghums are used for manufacturing a red dye and others as brooms. Maturation varies between 2½ and 7 months. Generally heavy soils are preferred and some varieties are especially suited to floodlands with clay soils. Several varieties have excellent drought resistance properties. Development of higher yielding hybrids has been delayed in the past by the poor germination of many varieties, susceptibility to insect damage and serious loss of harvest grain to birds. The pennisetum, pearl or bulrush millets (*Pennisetum typhoideum*), known in India as *bajra* or *cumbo*, are an important associate of sorghum varieties in traditional food crop systems in tropical Africa and southern Asia. The distributions of the two crops are probably very similar —data are as yet inadequate to establish the point—but significantly in India and West Africa the two crops commonly occur in crop mixtures or successions. Some pennisetum millets mature in barely two months and possess remarkable properties of drought resistance, producing satisfactory yields even on very light soils. Some awned varieties give excellent protection from the attacks of birds, often one of the most important causes of grain loss in the drier portions of the humid tropics.

Potatoes are the traditional root crop of the South American uplands just as yams are the traditional root crop of the equatorial lowlands of Africa and southeast Asia. Neither of these crops seems likely to make a major contribution to the food supply problem of the Third World in the immediate future. Environmentally both are restricted. Potatoes require a long cool and moist growing season and have become one of the most important foodstuffs of the temperate world, being restricted throughout the mainly tropical Third World to uplands, and finding little favour even in the sub-tropical margins. Yams are mainly of tropical African, especially West African, and Southeast Asian origin. Some varieties are now grown, especially for local commercial purposes, in moist areas in the savanna lands where cheap land or labour locally reduces costs, but generally the special requirements of yams of extremely moist conditions, long rainy season, deep loams, a high labour input in deep tillage, and special attention to the vines, make yam production very costly wherever cheaper alternative foodstuffs such as cassava are available or where development has resulted in rising standards of living and wages. No success in mechanizing yam production has yet been recorded and the crop, apart from its traditional importance in many communities, seems increasingly to be produced as a semi-luxury for higher income families.

As a local foodstuff groundnuts are universally grown in the humid tropics with a marked dry season. They are in most locations the chief source of cooking oil, their counterpart in the equatorial region being the oil palm. Light, even sandy soils, are preferred as the plant pushes the fruiting carpel into the earth. In West Africa, one of the world's most important areas of production, the soils of former Saharan dunes in Senegal, Mali, Northern Nigeria and Niger with a clay fraction of less than 10 per cent are excellent. In India the crop in its many varieties is particularly wide ranging, extending from the low rainfall and light soiled areas of the Punjab and Rajasthan in the northwest to the red loams and much wetter conditions of Madras in the south and Bihar and Orissa in the east. Varieties group into two major types: the running and the erect or bunched. The running varieties are the more productive, have a higher oil content and earlier maturity. They have a marked preference for light soils and chiefly occur in the lower rainfall areas. The erect varieties can grow on heavier soils and in wetter areas and are easier to harvest by machines.

Cotton is the most widespread fibre crop in the world and occurs in subtropical, warm temperate, and humid tropical areas with dry season conditions. A variety of species is cultivated but most of the world's cottons are derived from *Gossy-*

pium hirsutum (American Upland) and *G. barbadense* (Egyptian and Sea Island cotton). More and more cotton which was grown for export is today diverted to yarn mills in the producing country, partly because of the increasing market for cotton yarn in tropical countries and partly because of the increased use of synthetic fibres in the major consuming countries. Cotton can be grown by irrigation or on rainlands where low cost systems often find it economic to produce low yielding and short staple but hardy varieties. It needs a dry season for the ripening boll and is highly susceptible to a variety of diseases. Cotton is popular in many peasant farming systems not only because of its ready sale either on the local market or for export but because its distinctive times of planting and harvesting provide a better spread of labour utilization in the farming calendar.

In livestock production, which is mainly for local consumption of meat, milk, eggs, fibre and skins or leather, the Third World lags even more behind the more developed countries than in the production of crops. Not only are the numbers of livestock per unit of human population low, but the output of useful product per animal is also extremely low. Grigg comments that the tropics have half the world's cattle but produce only one-third the world's meat and one-fifth the world's milk (Grigg, 1970, p. 247). Protein shortage is an even more marked feature of Third World diets than shortage of energy foods and a more immediate and serious problem. Moreover, there has been much less demand for livestock as draught animals than in the mainly temperate more developed countries, excepting the widespread use of oxen and water-buffalo in southeast Asia. Ploughing still remains a poorly developed art in most of the world's savanna lands and hardly exists at all in the soils of the rainforests. A formidable array of diseases and pests affects livestock production in the humid tropics and veterinary services are still slight. The main problem, however, lies in the development of satisfactory tropical pastures. The potential feeding value of tropical grasses which grow rapidly and to a greater height than temperate grasses may seem considerable. In practice they hardly ever give a sward as in temperate countries which is satisfactory for herbivores, they rapidly become coarse providing high proportions of indigestible material and with rapid maturity the period of high protein content is short. Mineral deficiencies are frequent and in much of the tropics long dry periods of dried-up grasses of low nutritive value are normal and can be countered only by the use of stored foods, irrigation or migration to floodlands or areas with longer rainy seasons or with different seasonal incidence of rainfall. The necessity for low cost systems of production normally results in migration and as a consequence the separation of livestock production and crop farming.

The emphasis on low costs keeps livestock densities low except in India where the extensive use of draught animals combined with the religious veneration of cattle has combined to produce cattle over-population (probably a quarter of the world's cattle) in those areas most subject to human overcrowding. Outside India the most important areas of cattle and water buffalo concentration for local markets are in the Brazilian *sertão*, the Andean valley floodplains, Caribbean coastal lowlands and *llanos* of Colombia and Venezuela, the highlands of central Mexico, the nomadic pastures of the West African *sahel* and the drier nomadic lands of East Africa, the Ethiopian highlands and the rice growing lowlands of Southeast Asia (Fig. 17). Sheep, mainly hairy sheep, and goats attain large concentrations in northern India and Pakistan and in the southern Deccan, especially Madras. They are very important on the marginal grazing lands of the Middle East and on the extensively grazed *sahelian* and savanna lands of tropical Africa. In South America huge concentrations of wool sheep are achieved in the Argentine and Uruguayan pampas, in southern Brazil and in Patagonia which together provide major exports of mutton, lamb and wool. In the Andean uplands the unique ovine species, the llama and the alpaca are important, the former for carrying and the latter for wool. Pigs are more frequently than most livestock the subject of dietary taboos, especially in southern Asia and Africa. The largest concentrations, often closely associated with small-scale farming, are achieved in Brazil where the pig is both a scavenger and a major additional source of income to large numbers of small farmers. In Southeast Asia the pig again is important as a source of meat and cash income easy to raise on the waste products of agriculture. Poultry serve similar purposes, and since they are subject to fewer taboos and provide a more frequent and regular output in the form of eggs they have acquired increasing popularity throughout the Third World as one of the most important cheap sources of protein. The growth of the poultry industry is being encouraged in most Third World countries, often in enclosed, even battery systems, for which a regular supply of feeding stuffs, preferably maize, has to be secured. A major problem is that traditions of hand agriculture on small farms make the production of feeding stuffs costly or even unthinkable to peasants who reject the idea of growing crops to feed animals.

THIRD WORLD EXPORT CROPS

Raw materials, minerals and crops, have been the main exports of the Third World to the more developed countries, mostly in part processed forms for use in manufacturing or food preparation industries in Europe and North America. The most important export primary product items by value in the early 1970s have been oil, coffee, copper, sugar, cotton, oilseeds, cereals (maize, wheat and rice in that order) and rubber, each providing 3 per cent or more of total exports whilst tea, cocoa and meat provided no more than 1 per cent each (Bairoch, 1975, p. 100). Despite considerable progress in manufacturing in several less developed countries raw materials still account for approximately three-quarters of all Third World exports by value (one-third of all export earnings is derived from oil alone) whilst over two-thirds of the imports by value are manufactures. These raw material exports are concentrated into individual, less developed countries which tend to specialize, particularly in the production of certain crops. Bairoch calculated that in the Third World in 1963 55·6 per cent of all exports consisted of goods which were at the top of the list in individual countries compared with only 18·5 per cent in more developed countries (Bairoch, 1975, p. 101). In consequence diversification of trade is a major item of development policy in most of the Third World, whilst export crop distribution at the world scale shows patterns of extreme concentration. In part these patterns reflect crop sensitivity to environmental factors, but undoubtedly the most important factors have been the political and economic linkages between certain Third World countries and certain more developed countries mainly as a result of the imperialist and trading activities of the late nineteenth century when the basis of the present world trading system was largely created. One can hardly dissociate rubber production in Malaya, tea in India and Sri Lanka, sisal in East Africa, cocoa and oil palm produce in West Africa and meat in the Argentine and Uruguay from the development of the British Empire and sterling area, any more than one can dissociate coffee production in West Africa and Madagascar, sugar and bananas in Réunion, Martinique and Guadeloupe, groundnuts in Senegal and rice exports from Indochina from the French system.

The large trading systems that were developed in the 1920s and 1930s, buttressed by preferences, special pricing arrangements and most-favoured-nation agreements, have persisted despite new 'common markets' and trading areas and despite the General Agreement on Tariffs and Trade (GATT). In a sense the creation of the European Economic Community has simply widened the area of preference between the former imperial powers and their ex-colonies, which previously had only enjoyed preferences in the mother country. Such preferential trade has become an important issue of policy in the United States and in the excluded less developed countries, more especially in Latin America (Weintraub, 1966, pp. 16–17). Preferences are, however, supported not only by more developed countries with an important position in world trade but by less developed countries seeking special or virtually guaranteed markets for their produce. They are still a major factor in world export crop distribution and were the cause of a major division in attitude between the Latin American countries, who feel discriminated against, and the African countries, who believe they benefit from them, at the first United Nations Conference on Trade and Development (UNCTAD) in 1964 (Weintraub, 1966, pp. 83–8). These international arrangements, including special trading arrangements between groups of countries within the Third World such as the Andean Common Market, the Central American Common Market and the East African Community, might be regarded as providing regional units of significance for agriculture at a scale between that of the world and of the nation. Their geographical effect is, however, as yet not easy to determine in relation to Third World export crops. More marked have been the effects of international commodity organizations, such as the International Coffee Organization, but these differ for each commodity, tend to preserve geographical distributions already established rather than create new patterns, seem to maintain inefficiency where they work at all and have been frequently subject to failure. Thus the Diversification Fund set up under the International Coffee Agreement of 1968, the first commodity agreement to establish obligatory production goals, was wound up after five years through poor fund raising in the consuming countries and improved world prices and prospects for coffee (R. Watts, 1973).

The price elasticities of world demand for most Third World agricultural products tend to be low and the prospects for greatly increased sales of many commodities such as coffee, cocoa and sugar would seem limited. The brightest prospects are for timber, fish and meat products (Helleiner, 1972, p. 40) rather than for the traditional export staples, more especially beverages, oilseeds and fibres. 'Only the remarkable and sustained rates of growth of the industrial nations were able to maintain the modest rates of growth of agricultural exports from poor nations in the 1960s' (Helleiner, 1972, p. 35). Moreover, during the 1950s and 1960s the terms of trade were especially adverse for the Third World (Fig. 18). The agricultural commodities exported bought smaller and smaller quantities of industrial goods, thwarting development plans made in the late 1950s and early 1960s and making the creation of an infrastructure for the improvement of agricultural productivity difficult or impossible. This deterioration in the terms of trade, which has discouraged hope of development through export cropping, did not exist before 1926–9, when the secular changes of price favoured primary products, and has been a marked feature mainly of the period 1952–62 (Bairoch, 1975, pp. 111–34). Streeten (1974) has argued that the terms of trade for most less developed countries were unusually favourable in the early 1950s and that it is difficult to show any long-term trend, that there is no evidence of a secular deterioration of the terms of trade of primary products in relation to manufactured products unless one chooses base years arbitrarily in order to prove the their relation to the market and to other crops in the agricultural system.

The Third World produces most of the world's coffee, totalling 4·5 million tonnes in 1975. 3·4 million tonnes were exported as green roast—97 per cent from the Third World. In a more 'normal' trading year, 32 per cent came from Brazil and 11 per cent from Colombia. Other important concentrations of coffee for export production are in Central America, the Ivory Coast, Angola and Uganda (Fig. 19). Significant minor centres of production are in the West Indies, Venezuela, Ecuador, Peru, Equatorial Africa, Kenya, Ethiopia, Yemen, southern India, Laos, Vietnam, the Philippines and Indonesia, especially Java and Sumatra. Coffee is widespread and yet most of its production is confined to a very few countries and mainly either to tropical uplands with well drained yet moisture retentive, slightly acid soils suitable for the higher quality *Coffea arabica* varieties, or to equatorial lowlands where the coarser *Coffea robusta* varieties especially suited for the manufacture of 'instant' coffee are grown. Rainforest coffees can grow in shade but its necessity is doubtful, although light shade from *Erythrina* spp. or *Albizzia* spp. is favoured in the Ivory Coast. Formerly the market was dominated by *arabica* coffees especially those from Latin America, but more recently African producers have made big gains in the world market with *robusta* coffees, especially in Uganda, Angola and the Ivory Coast. In some respects the distribution pattern seems curious. One might have expected the crop to be more widely distributed, and many favoured areas are without it. Part of the answer lies in the evolution of world trade and associated political and economic linkages between more developed countries and the producers, and part lies in the complex story of the diffusion of coffee plants and their important association for a time with white settlement in the tropics and the development of plantations. *Robusta* expansion came late in the nineteenth century. *Arabica* development dates mainly from the early eighteenth century and the importance then of the Portuguese world trading system and empire, and their concentration of effort on the production of tropical export crops in Brazil. As early as 1690 the plant had been introduced into Java. Southeast Asia and Sri Lanka became for a time the world's leading producers of high quality coffees until most of the plantations were destroyed by fungus diseases, especially the leaf disease, rust, caused by the fungus *Hemileia vastatrix*, for which there is still no satisfactory method of control. Coffee and empire came late to tropical Africa and would hardly have succeeded in West Africa without special French encouragement, including for a time heavily subsidized prices in the French market (about 64 per cent above world prices in 1960, for example). Despite the dominance of Brazil and the importance there of plantations, most coffee production comes from smallholdings and has brought greatly increased incomes to former peasant farmers. Many have achieved remarkable savings and levels of investment in order to plant the many millions of trees involved, to acquire land for tree planting, to wait 6 to 8 years for the trees to come into bearing, to hire labour, to learn to prepare the crop for market and, in some cases, to build roads and bridges and form co-operatives to sell the crop and purchase pesticides and depulping, fermentation and drying equipment. As a tree crop coffee means the 'permanent' or rather long-term occupation of land, which in many African and some Southeast Asian countries with usufructuary systems of land tenure has meant a revolution in landholding and settlement (see pp. 139–40). The rapid expansion of the crop has also meant a migration of white settlers and of local farmers and the creation of a coffee frontier. This is advancing until all available coffee land in the main producing regions will have been taken and the consolidation and intensification, together with more diversified forms of farming, which have already begun, will become universal. One final point in the world coffee map is the importance of the United States as a consumer, taking nearly half the world's exports, and the special relationship in trade with the Third World which the United States has sought in Latin America (Weintraub, 1966, pp. 38–43). This has resulted in trade reciprocity treaties with those Latin American countries which, excluding Chile, Argentina and Uruguay, were not concerned to maintain trade ties with Europe.

Sugar cane is a universal crop of the humid tropics and subtropics. Of the tonnage of sugar exported 71 per cent was from Third World countries in 1974, including 51 per cent from Latin America. Mostly it is cultivated either under irrigation or in areas with a rainfall generally over 1200 mm. As an export crop of the Third World its distribution, although in a sense widespread, is surprisingly restricted, being mainly confined in order of production to Cuba, Brazil, the Philippines, the Dominican Republic, Mauritius, Mexico, Argentina, Peru, Fiji, Thailand, Guyana, Colombia, India and Réunion (Fig. 19). Hawaii, Taiwan, Jamaica and Madagascar are also sugar exporters of some importance. Huge quantities are produced in South and Southeast Asia, Egypt, tropical Africa and most other Latin American countries, but mainly for home markets. Sugar cane is of tropical origin, but its distribution has spread beyond the tropics and the Third World as new varieties have been developed, so that today the United States, for example, is a major grower. Most temperate countries plant sugar beet, which provides over 40 per cent of the world's centrifugal sugar supply. Sugar cane was a common crop on peasant farms in Asia and Africa where the product was mainly chewed. The export of refined sugar to Europe led to the development of the world's first tropical plantations for the overseas trade using African slave labour. These were established first of all on West African islands, beginning in Madeira in 1420 and arriving in the Americas at San Domingo by 1494. By 1520 Príncipe was a source of sugar for Europe, followed by Sao Tomé and Fernando Po and shortly afterwards by northeastern Brazil, where the import of African slave labour and the export of sugar laid the foundations of a trade network which developed into the 'South Atlantic System' (Curtin, 1965, p. 4) between the West Indies, Europe and Africa in the eighteenth century. Thus the concentration of sugar production in islands and in Latin America has its foundations in slavery, plantations and nearly five centuries of trading history. Few tropical African countries have become major producers. Most of them are still sugar importers although most have attempted to develop plantations, usually with foreign capital and advice. The persistence of the plantation or of estates in which small farm production is centrally controlled, despite the ease with which peasant farmers may grow the crop, derives from the importance of the refining process, the tendency to increasing size of sugar mills and the need to control the production and supply of raw materials for economic operation. A large mill can process the crop from 2000 to 4000 hectares in a season. Modern commercial sugar canes are hybrid, *thick* or *noble* canes bred from five species for disease resistance and a good nitrogen response, and requiring 8 to 24 months to reach maturity. The *thin* canes which can tolerate lower rainfall and poorer soils are mainly grown for fodder. Well-drained soils high in organic matter, particularly silty loams, are preferred for commercial canes, but reddish brown lateritic soils in Hawaii low in mineral status nevertheless are productive with heavy nitrate dressings and the correction of manganese and boron deficiency. *Noble* sugar canes are particularly susceptible to mineral deficiency diseases, which are widespread and need careful control. They are also susceptible to storm damage and many of the major sugar producers are visited by hurricanes or by other tropical revolving storms. Part of the concentration is to be explained by preferential trading arrangements between the United States and Cuba (until the Cuban revolution in 1958–9) and the Philippines, for example, or between Great Britain and the former British West Indies.

The general distribution of cotton cultivation has already been discussed, for cotton is mostly used for yarn production in the country where it is planted, and the expansion of cotton textile industries is one of the more marked features of industrial development in the Third World. Nevertheless there is still much cultivation for the export of cleaned lint and seed. Of the raw cotton tonnage sold abroad in 1974 the Third World countries exported 49 per cent including 21 per cent from Middle East countries and 17 per cent from Latin America. In 1973 Brazil was the largest Third World exporter by weight followed by Egypt, Sudan, Pakistan, Mexico and Syria, but Egypt was easily the leading Third World cotton exporter by value (12 per cent of world exports to Brazil's 5 per cent) because of the concentration of its farmers on high quality *G. barbadense* varieties. Although cotton may be grown on a wide range of soils, it does best where soils are well drained and aerated yet retentive of moisture. Susceptibility to disease in conditions of high humidity plus the need for dry weather when the crop is ripe make cotton a risky crop of highly variable returns. The advantages of irrigated production are considerable, especially for the higher quality cottons, and risks are greatly reduced. The strength of

United States production of medium and short staple cottons and widespread competition in the world market have resulted very often in low prices for the poorer qualities mainly produced by peasant growers on rainlands in the Third World. For farmers in tropical Africa and Brazil export cotton cultivation has had very varying fortunes and has normally been less profitable than several other crops, but the crop has nevertheless remained attractive partly because of its place in the farm calendar, partly because it is a field crop normally treated as an annual giving some flexibility in production, and partly because in the humid tropical rainlands with a marked dry season the range of export crop possibilities is often quite limited. Only in South and Southeast Asia where so much of the population is concentrated in floodlands and dependent mainly on rice cultivation is export cotton production somewhat unimportant and even here cotton is still a significant crop for the home market.

Rubber production is almost entirely confined to plantations and smallholdings of *Hevea brasiliensis* or Para rubber, a native of the equatorial rainforest. It is hardly surprising that 98 per cent of exports by weight came from the Third World in 1974, but with 92 per cent from Southeast Asia the regional concentration of production was nevertheless startling (Malaya 46 per cent, Indonesia 26 per cent). The minor production centres which are of some importance in the world market include Sri Lanka, Liberia and Nigeria. Huge areas suitable for production exist in Africa and Latin America, where Brazil at one time exercised a world monopoly of wild Para rubber export, its only competition being from the inferior product of other wild plants, notably tropical vines. The demand for rubber on a scale to warrant rapid expansion came in the late nineteenth century when the most accessible areas of equatorial rainforest which were controlled by European powers were in Southeast Asia. The pre-eminence in production of the coastal plains of western Malaya, closely associated with the dominance of Singapore in the trade of the Far East, made this former British colony the leading choice for plantation enterprise in rubber production, closely followed by neighbouring islands in Indonesia and by Thailand. The African rainforests were less accessible, as in the Zaire Basin, despite river transport, or were committed, like southern Nigeria, to fostering peasant farming. In Brazil production was dominated by the gatherers of wild rubber, who persisted despite high costs. Major plantation investment was directed to other crops and, perhaps more significantly, to other regions nearer the centres of political and financial power. Plantations were dominant in the early stages of the expansion of rubber production mainly because of the high costs involved—highest yields came from double-budded trees—and the need for careful control of tapping. Today, often with government assistance or through growers' co-operatives the special techniques and high quality plant stock needed can be acquired, and smallholding rubber production is increasing much more rapidly than plantation production.

The main sources of vegetable oil of commerce in order of approximate importance by weight as Third World exports are groundnuts (peanuts), coconuts, soya beans, oil palm fruit and kernels, cotton seed, linseed, sesame, sunflower seed, castor and tung. It is virtually impossible to calculate the exact order of importance by weight of material exported, because this is in varied forms and depends on different extraction methods. Moreover, the oils find markets both as food and as industrial materials, and the plant by-products include animal feeding stuffs. The chief exporter in total value of groundnuts, oil and groundnut cake and meal is Nigeria, still a large exporter of nuts, followed by India, exporting mainly cake and meal, Senegal exporting chiefly oil and cake, Brazil, Argentina and the Sudan. Most exports are from relatively dry, light-soiled areas where peasant growers on smallholdings have little choice of cash crop and cultivate groundnuts as a cash supplement to the subsistence food crops. Northern Nigerian production depends on especially favourable railway freight rates and still includes a considerable export of nuts in shell. Senegal, until recently, was the world's leading oil and cake exporter, but production has declined sharply in recent years because of the ending after France joined the European Economic Community of the special French prices, which were well above those of the world market and which led growers to average about two-thirds of their holdings in groundnuts and depend for their food staple on imported rice. Like the groundnut the coconut palm is grown mainly by smallholders, the main centres of export production of copra, oil cake and meal being the Philippines (with 60 per cent by value of world export production, chiefly in the form of copra and oil), Papua New Guinea, Indonesia, Moçambique and Sabah. The tree grows so well on light sandy soils in coastal situations where little else is comparably productive that its concentration as a commercial crop in the islands and peninsulas of Southeast Asia is hardly surprising. Other tropical coastlands seem remarkably lacking in development, often producing coconuts virtually as a subsistence or local market crop. Soya beans are probably of warm temperate origin but varieties extend from cold temperate to equatorial latitudes, more especially on the eastern sides of continents where in temperate latitudes the high summer temperatures are favourable. The Third World exports only 17 per cent of the total value of world trade in soya beans, oil cake and meal, but the crop is of increasing importance for less developed economies especially in Brazil, Paraguay and Indonesia. The enormous environmental range of its varieties, however, suggests that many countries will attempt to develop their own production and that the oil produced by them will increasingly be used as a substitute for other Third World vegetable oils. The oil palm belongs mainly to the continuously moist humid tropics and export production comes mainly from Southeast Asian, Central African and West African plantations and smallholdings. Plantation palm oil, mainly from Southeast Asia and Central Africa, is generally of higher quality (lower free fatty acid content) but modern margarine and cooking fat technology is making that feature of much less importance. The trade is in palm nuts and kernels, palm oil, palm kernel oil and palm kernel cake meal, of which the main exporters are Malaya, Indonesia, the Ivory Coast, Zaire, Nigeria and Benin Republic.

Third World cereal exports total led some 18 million tonnes in 1973, against imports of some 48 million tonnes. Whilst overall, therefore, the Third World is a cereal importer, nevertheless the export, mainly of maize, wheat and rice, is of special importance in the agricultural economies of a few Third World countries. Third World maize exports are mostly from Argentina, Thailand and Brazil. Argentina's production is concentrated within 200 miles of the coast and is exported mainly to Europe and chiefly for poultry feed. Growers have concentrated on the small-kernelled varieties of flint corn especially suited for the purpose and not on the dent varieties for fattening hogs and cattle, for which they prefer alfalfa. In effect the Maize District of the Humid Pampa of Argentina has been created especially for the export trade to Europe. Brazil's maize exports have developed more slowly and recently, as farmers, especially those in the south already growing maize to fatten cattle and pigs, have been able to increase production sufficiently to produce a surplus above local needs. In Thailand maize production has expanded since 1945, chiefly for use as animal feed. The maize is grown chiefly in the great rice-growing district of the flood-plains of Central Thailand, where farms above average size for Southeast Asia have improved yields with new irrigation facilities and crop plant varieties and have made possible considerable food crop surpluses, permitting the diversion of resources away from local food grains. Third World wheat exports are chiefly from sub-tropical and warm temperate lands in Latin America, more especially from the Humid Pampa of Argentina and from northwestern Mexico. Here again circumstances differ. Argentina's export of wheat from the Humid Pampa is old established and based on the early development of an export trade to Europe and the significance of wheat as a major cash crop in short-term tenancies designed to prepare land for the planting of alfalfa. Mexico was not independent of imported wheat until 1956 after the introduction of new high-yielding wheat varieties from 1948 onwards, associated in most areas, especially in Sonora, the Imperial Valley and the Laguna District, with irrigated lowlands. The chief rice exporting countries of the Third World are in Southeast Asia, particularly in the not so densely occupied irrigated lowlands of Thailand, Burma, Khmer Republic and South Vietnam. Other important exporters, however, include Egypt, which exports high-value rice to pay for lower-value food grains, Taiwan and Pakistan, which has achieved high levels of production by the introduction of new high-yielding rice varieties in the 'Green Revolution'. Most production for export in the Third World comes from larger than average smallholdings, able in recent years to achieve high yields from new varieties and fertilizers or in the past to achieve high outputs per man by more extensive methods. Both in Taiwan and in Thailand the achievement of an export surplus has been subject in part to government price and marketing controls.

THIRD WORLD AGRICULTURAL INNOVATION

Changes in crops, inputs, marketing, equipment and agricultural techniques have already been discussed in some detail at the world scale, inevitably as part of the explanation for the current situation and as a feature of present-day agriculture. Few situations are static and decisions by farmers regarding their crops and their agricultural systems must refer at least to the immediate future and past. For the most part innovations as such are best treated at the local and regional levels, with reference at the national level to national planning and government as an innovation factor. Nevertheless, there are some world trends and even some attempts to plan, in very broad terms, at the world scale.

In general, crop innovation in the Third World has occurred less frequently than elsewhere and has diffused more slowly. This is hardly surprising in societies whose main crops were largely for subsistence and grown in systems which had achieved some measure of adjustment to the limitations of local labour supplies, environment and seasonal rhythms. In many cases the introduction of new higher yielding varieties either offered a surplus for which there was no market or a labour gain only at times of the year when labour had little alternative work to do. A gain in leisure is usually appreciated, but has to be considerable to offset the risks of change. Most of the changes in cropping that have taken place in the peasant sector of production have been for the introduction of some preferred taste or of a crop which offers surer returns in situations of environmental variability or fits better into the cropping calendar. Many ancient crop plants have almost disappeared or are known today only in their wild forms or as weeds, whilst their place has been taken by the now universal wheat, rice, maize, sorghum and cassava. Thus in tropical Africa, cultivated *Brachiaria deflexa* (*founi kouli*) is hardly known outside the Fouta Djallon of West Africa although widespread as a wild edible grain; the once important earth peas (*Voandzeia geocarpa* and *Kerstingiella geocarpa*) have been largely replaced by the groundnut (see earlier references, pp. 30 and 62); cultivated Digitaria grains have given way to *Eleusine coracana* or finger millet (Johnston, 1958, p. 60); and in South and Southeast Asia cultivated species of *Panicum* are today minor cereals more commonly found as catchcrops on poor soils. Whilst some of this diffusion has been shown to take place by direct contact between agricultural communities, as, for example, in the spread of maize across tropical Africa after it had been introduced to the Old World by Europeans returning from the Americas (Portères, 1955), most of it was the work of the seamen, traders, migrant farmers, amateur gardeners and agricultural scientists of the great empires which have been created in the Third World from the period of the Arab conquests to the present day.

Most farm businesses in the Third World are minute. They have little or no spare labour, space or capital to experiment with new crops or techniques, or to test the claims of would-be innovators or seeds merchants. It is hardly surprising that most crop innovation has come as an extension of the existing work in crop innovation already active in the more developed countries and has been most rapidly diffused amongst farmers producing for the overseas market, or at least who are in the commercial sector and hope to be able to market their increased production. Most work in crop improvement has been amongst crops, such as maize, already important outside the Third World, or amongst major commercial export crops, such as coffee, cocoa and rubber. Much of the experimental work has been either in foreign-owned plantations or in the experimental farms created by government agricultural departments, which at least before 1950 were largely staffed by ex-patriate agricultural scientists.

In this century agricultural innovation in the Third World whether of crops, equipment inputs or techniques has been largely, although not entirely, the work of government, usually colonial, or of foreign or foreign-owned interests. For much of Africa, Asia, Oceania, parts of Central America and the West Indies its geographical spread was therefore largely governed by the distribution of the empires and their rival economic and research systems. Political considerations were of great importance in the operation of the networks of research stations and demonstration farms as instruments of social and political change, and the effect, although small in relation to the whole and slow in its spread, was nevertheless considerable in relation to the small numbers of scientists and their co-workers involved (Morgan, 1973, pp. 8–9). For example, the entire British Colonial Agricultural Service was staffed with hardly a thousand graduate officers in thirty years yet bred many new varieties of major economic plants, introduced many thousands of crop plant varieties, created research systems and extension services, identified many pests and diseases and introduced methods of cure or control, created veterinary services, undertook major work in plant ecology and in studies of soil fertility, crop combination and rotation, introduced equipment of all kinds together with fertilizers and sprays, pioneered management systems and accountancy methods for peasant farmers engaged in commercial cultivation for the first time, investigated and tried to resolve problems of land tenure, and helped to improve marketing and produce quality (Masefield, 1972, pp. 4–5 and 76–104). Not all large plantation or estate systems favoured innovation. In Hawaii and Malaysia, for example, large-scale foreign-owned interests were largely responsible for the introduction of improved plant stock and techniques for the cultivation of sugar cane and rubber, but in Latin America, whilst the *latifundia* were ready to adopt the latest boom crop and plant it on an expanding frontier of production, many of them were less ready to experiment with new varieties or to use fertilizers or sprays. The greatest profits were often achieved more by reducing inputs to the lowest level possible, even if this meant lower yields. Only the disappearance of the frontier and rising land costs could encourage greater inputs and more careful husbandry.

Diffusion has also been conditioned by the character of the innovations themselves. For much of this century not only has a large part of the research in Third World agriculture concentrated on export crops, it has also concentrated on particular aspects of export crop production, more especially on the quality of the product and on yields. Where the market has been sensitive to quality and where land shortage has occurred or competition for favourable locations has resulted in rising land costs these emphases in research have been largely successful. But in many cases the production problem was labour shortage or was the result of the role of a particular crop in a farming system involving other enterprises, or was part of a production and marketing situation in which the key element was highly uncertain and fluctuated in price. Thus in some areas innovations were rejected and attempts were even made to force them on peasant farmers by law so that crop refuse had to be burned by certain dates in order to control pests, box ridging had to be undertaken and the rules of 'good husbandry' observed. Paternal government interference of this kind, however, has rarely succeeded. As Allan has noted: 'There is little reason to suppose, and no experience to suggest, that the principles of farming can be imposed upon a society of semi-subsistence cultivators by government order, and it is unlikely that legislation can destroy the kinship system and remove the burden it places on successful members of the group . . . Efforts . . . to impose the good husbandry regulations against a prevalent apathy and some degree of hostility, and at the same time to maintain the records of a constantly changing population-land pattern, may well create an insupportable administrative burden' (Allan, 1965, pp. 426–7). Apathetic farmers, especially if 'semi-subsistent', are frequently not interested in improvement because the efforts involved seem out of proportion to the expected rewards. Major breakthroughs in commercial production, such as the introduction of cocoa-farming in Ghana, offered rewards which seemed enormous to peasant farmers and which locally resulted in an agricultural revolution, so rapid and effective were the changes involved. As export markets expand less rapidly, however, and the taxes on export crops increase, the rewards may seem less and the prospects for innovation in local food production may also not appear particularly attractive. Never has the business of agricultural innovation been so active and widespread in the Third World, but with the difficulties of expanding both internal food and export markets the current prospects in many countries are less encouraging for investment.

Despite the difficulties, many Third World governments and international agencies have been especially concerned with increasing staple food production, both to improve local nourishment and to reduce in certain instances costly food imports. The Green Revolution was, and in effect still is, a world wide attempt to revolutionize the production of wheat and rice in several Third World countries. Its techniques have also spread to certain other cereal crops which have locally achieved some more limited success. The new seeds are hybrids or high yielding varieties (HYVs) of wheat, initially bred in the International Maize and Wheat Improvement Centre in Mexico in the 1950s, and of rice (especially the IR-8 variety or so-called 'miracle rice') bred in the International Rice Research Institute (IRRI) in the Philippines in the 1960s (Johnson, 1972). These new hybrids have proved highly responsive to fertilizers in conditions of adequate water supply and effective husbandry. Spectacular yield increases have been achieved, even more than double the yields of traditional varieties, paying not only the cost of the increased inputs but providing considerable increases in farmers' incomes.

The new wheat seeds made possible a three-fold increase in wheat production in Mexico in twenty-three years (1944–67) and laid the basis for accelerated developments elsewhere, notably in Pakistan, where 42 000 tonnes of the new seeds were imported in 1967–8 and the area under the HYVs increased from 4900 hectares in 1966 to 2·4 million hectares in 1969. By that same year in South and Southeast Asia nearly 14 million hectares were planted to new varieties of both wheat and rice (Brown, 1970, pp. 3 and 19–21). In India by 1971 after barely nine years of effort the HYVs accounted for over a third of total grain production (rice, wheat, maize, jowar and bajra) (Jayaprakash, 1973). There between 1965 and 1973 over 2500 lines of dwarf rice were generated and evaluated and 48 varieties, some introduced and some local, have been released for cultivation. By 1973 India was already using the second generation of HYVs with new more disease resistant varieties replacing earlier introductions. Several of the dwarf rices mature very quickly and fit well into multiple cropping patterns (Ram, 1975). By 1968 the Philippines and Pakistan had attained self sufficiency in rice production and so rapid were the subsequent gains in both wheat and rice production that by 1972 an FAO report recommended a cutback in wheat production in the more developed countries in order to maintain market equilibrium and a 'land diversion' programme for rice (FAO, 1972, pp. 20–1), but in 1972 the monsoon rains failed in most of southern Asia and production of foodgrains, instead of increasing, dropped by 3 per cent. By early 1973 India was buying wheat on the world market and had begun relief operations. The HYVs, however, have shown that enormous production possibilities can be realized. Geographically they are location selective since the best results are achieved where cheap irrigation makes possible satisfactory water control. There is also some geographical effect from the distribution pattern of research co-operation and its relationship to the central office or control centre, as for example in the geography of IRRI co-operating research agencies and trainee sources in relation to the Institute at Los Baños in the Philippines (Fig. 20 based on data in Barker, 1970). They also are most successful where overcrowding encourages increased intensity of production. The new seeds are a substitute for both land and labour since the productivity of both is increased, but more especially they raise yields and are a substitute for land. Much less has been achieved in the humid tropics, particularly in the low population density areas of Latin America and Africa. The most successful achievements have been in Mexico, India, the Philippines, Pakistan, Bangladesh, Thailand, Burma, Indonesia, Malaysia, Sri Lanka, Iran, Afghanistan and Turkey. Within the countries affected the Green Revolution is location selective, and once the market for wheat or rice is saturated it encourages a concentration of production in particularly favoured localities. The commercial and environmental risks are raised with increased dependence on success in the market to pay high costs and increased dependence on a satisfactory water supply. An inevitable result with so fast a development has been a boom in irrigation equipment and water control schemes, a rise in fertilizer consumption of 13·8 per cent a year from 1961–3 to 1968–70 (FAO, 1974A, p. 28) with consequent considerable price rises, similar rises in demand and price for pesticides and herbicides, increased dangers from disease and therefore increased dependence on the plant breeding organizations to produce newer crop varieties fast enough to keep ahead of the spread of new diseases, bottlenecks in labour supply more especially at harvest time, and a widening income gap between

the richer and more successful farmers, better able to take advantage of the new seeds, and the poorer farmers. Serious bottlenecks have also occurred in distribution and trade, in the supply of equipment such as drying facilities, and in storage. Not all the new crop varieties have found a ready market. Some of the new rice varieties have proved unpopular because of coarseness of grain and stickiness in cooking (Jayaprakash, 1973). There is also some evidence that the new cereals have a lower protein content than the older, lower yielding varieties and are spreading at the expense of the area in legumes (Altschul and Rosenfield, 1970).

Although no one at the world scale can decide the geography of world agriculture, nevertheless a large number of international organizations and their agencies are influenced by FAO, which provides an annual monitoring service for world agricultural performance together with analyses, reports and recommendations. On the basis of this and other information, mainly from national government sources and from a number of very large international corporations, especially those supplying agricultural chemicals, fertilizers and machinery, decisions are made by international and national bodies concerning the supply of capital, trade regulations, international exchange rates, volume of trade, research and the supply of new seed, chemicals and equipment, which do have important effects on the world distribution of agricultural activity. Admittedly the services and financial aid they provide, whether in the form of grants or loans, is often only a small contribution to agricultural improvement. The research organizations sponsored by the United Nations and other bodies have as yet, with the exceptions discussed above, made only a small advance in an enormous field of low productivity and general poverty. Nevertheless, the influence of international evidence and debate whether in the United Nations or elsewhere has undoubtedly had a profound effect on the policies of many governments and on the regulation of world trade in agricultural commodities. The existence of GATT, UNCTAD, the International Monetary Fund (IMF) and the various international commodity organizations have not been without effect despite their failures. The UNCTAD committees, for example, review marketing and distribution systems, commodity diversifications and debt problems. They also recommend reserve stock levels and warn of critical problems and dangers in world trade.

Aid in the form of grants and loans to agriculture or to agricultural services has generally been less spectacular than aid to other sectors of the economy, although the urgency of the agricultural problem and some aid failure in other sectors have encouraged increased attention to agriculture in recent years. Some indication of the research expenditure problem is given in Table 2. This urgency has been increased by the realization that it is those Third World countries which are not oil pro-

TABLE 2

A COMPARISON OF RESEARCH AND EXTENSION WORK EXPENDITURES IN 1965 IN THE DEVELOPED AND LESS DEVELOPED COUNTRIES

	Expenditures on research million dollars	*Scientist man-years*	*Expenditures on extension million dollars*	*Number of extension workers*	*Expenditure per farm $*	
					Research	*Extension*
Developed countries	985·7	49 262	559·2	87 428	17·25	9·78
Less developed countries	126·6	10 298	140·9	76 412	1·07	1·19

Source: Evenson and Kislev, 1975.

ducers and which are mainly dependent on agriculture that have borne the brunt of the economic difficulties of 1973–5 and that have the fastest rising trade deficits. The chief sources of such grants and loans are the United Nations Development Programme (UNDP), the World Bank or International Bank for Reconstruction and Development (IBRD), the International Development Association (IDA), the International Finance Corporation (IFC), the Organization for Economic Co-operation and Development (OECD) and its Development Committee (DAC), large private foundations such as Kellogg, Ford, and Rockefeller, national governments, more especially those of the United States and the USSR, and their agencies, such as the US Agency for International Development and the US Development Association, the West German Deutsche Entwicklungsgesellschaft, the Dutch Financierungs Maatschappij Voor Outwickenlingstanden, the French Compagnie Financière pour l'Outre-Mer and the British Commonwealth Development Corporation and Commonwealth Development Finance Company, together with regional organizations, such as the European Economic Community, and their agencies. The receiving countries also possess regional organizations through which aid may be channelled, as, for example, the Colombo Plan for Co-operative Development in South and Southeast Asia, set up in 1950, the Special Commonwealth African Assistance Plan and the Economic Commissions for Africa, for Asia and the Far East and for Latin America. Organizations such as the Committee for Technical Co-operation in Africa (CCTA) and the Scientific Council for Africa (CSA) distribute no funds but co-ordinate scientific and technical activities and provide scientific advice for the African countries. In addition there are considerable private investment and loan arrangements which affect agriculture in the Third World, much of it in effect noted already in the discussion of plantations. More recently the huge monetary surpluses which have been amassed by some of the oil-producing states have in small part become a source of aid or investment. A major problem, however, is that world inflation, more especially in 1974, 1975 and 1976, is causing an actual diminution of aid, partly by the lowered value of the grants and loans and partly by growing reluctance on the part of donors to increase aid to compensate for inflation or even in some cases to maintain it at current levels. The United Nations has an aid target of 0·7 per cent of the GNP of donor nations. In 1974 the combined aid of the leading 17 donors was only 0·33 per cent of GNP. The total volume of $32 000 millions in 1974 was only a minute contribution to the capital shortage problem of the Third World and has failed to increase productivity by any significant amount.

Generally IBRD money finances large-scale schemes, particularly major irrigation and flood control schemes, provides for agricultural credit, settlement projects, large schemes for expanding treecrop production and livestock farming and related services such as rural health and education, water supply, agricultural feeder roads, railways, harbours and power supply. Between 1969 and 1974 the World Bank doubled its spending on agriculture and concentrated much of its expenditure in projects aimed at the poorest areas (generally those inhabited by people with an annual income of $50 or less) and more especially at raising the productivity of small farmers. These projects required for the most part reform of both land tenure and fiscal policies. UNDP funds have been directed to an enormous variety of agricultural projects and depend for their allocation on equal sums being provided by the governments of the less-developed countries receiving the aid. IFC funds are for productive private enterprise and are directed mainly to manufacture and mining, but include fertilizer production, food processing and canning, textiles, pulp and paper manufacture and leather tanneries. Most aid has major political implications and the largest flows have been to countries which were at the centre of the struggles between rival great powers or even engaged in war. The United States has been the largest source and some of the biggest flows have been to South Korea and South Vietnam. In the latter the effect of aid was to increase imports especially of basic commodities such as rice, thereby decreasing interest in rice production and encouraging a switch to the more lucrative vegetable crops. The vast purchasing power pumped into the economy by the war resulted in a grossly overvalued currency, which made it almost impossible to develop agricultural exports, encouraged capital flight and deterred foreign investment (Myint, 1972, pp. 116–26). The flight of the rural population to the towns was encouraged or to an urban market-oriented agriculture, a trend which the new communist government is concerned to reverse with the complete alteration of trading relationships as a result of the war. Undoubtedly the Latin American countries have attracted an enormous share of United States investment, both public and private, so that Latin American agriculture has been more affected by foreign investment, loans and grants than that of any other comparable area in the Third World and is also generally much more oriented towards exports.

3

The national scale

The nation provides the spatial framework for one of the most important decision-making bodies in agriculture—government. In some cases government makes locational decisions with regard to crops and livestock. It is often a major land and farm owner, with state farms not only contributing to agricultural research but providing a large share of the national farm output. Generally, however, the farmers are still 'free' to make locational decisions, but are constrained in these by government policy. This acts through the manipulation of marketing and pricing systems, grading, control of transport and storage, controls on dealers, manufacturers and processors, provision of services to agriculture, laws and regulations on almost every aspect of agricultural production and provision and maintenance of the social and economic rural infrastructure. Most governments, even in the poorest of Third World countries, attempt from time to time to measure agricultural productivity and trade and to formulate policy for agriculture. Agriculture in most Third World countries is the largest sector of production, the biggest contributor to gross national product and in consequence the chief target for taxation to pay for social and other services and to pay for government development projects, whether in agriculture or in other sectors of the economy.

The national economy may be regarded as a single, spatially bounded, economic system, of which agriculture is a major part. Agriculture occupies most of the national space, and has to compete for scarce resources with other sectors of the economy. It depends on services and more especially on a national trading and marketing system with a considerable regional variation in quality, and containing its own variations in the geographical distribution of its enterprises and systems of production. These result in a specialization marked enough in its locational character to qualify for the application of the term 'agricultural region'. Third World nations contain an enormous variety of agricultural distribution patterns, many of which are far from clearly defined even where data exist, but for most of which the data are in any case too imperfect to justify anything more than extremely limited conclusions about the nature of the distribution. There are further problems in trying to understand the geography of the national agricultural structure in that the distributions and regions concerned may be seen at different scale levels. These range from broad zones which may reflect the physical and biological factors of agriculture combined with demand and with the degree of appreciation of those factors by farmers, research workers, dealers and government, to the nodes or local concentrations of particular enterprises. The latter tend to reflect local variations in either production or demand conditions and may relate strongly to the system of central places. The spatial structure of the agricultural economy, just like the national economy of which it is a part, 'is not only the geographical manifestation of a particular type of economic and social development, but also an important compound in helping to determine the success or otherwise of development efforts . . .' (Odell, 1974).

Some observers have developed the argument concerning the relationship between regional variation in economic growth and stage of development. Williamson (1965) has attempted to order the apparent variety of regional patterns in development and income by invoking the relationship between regional dualism, which is most marked in agriculture and in nations whose technology is localized by regional resource endowment, and national income. He has argued that there is evidence of a consistent relationship between rising regional income disparities and increasing 'North-South' dualism, which is typical of 'early development stages', while regional convergence and a disappearance of severe 'North-South' problems is typical of 'the more mature stages' of national growth and development. The consistency of this relationship has not, however, been equally evident to all researchers, and no mechanism has been suggested by which regional income differentials first increase, then stabilize and finally diminish in 'mature' periods of growth. Moreover, much of the argument depends on a comparison of countries at different supposed stages of development rather than on a study of development itself. In places a spatial comparison is made to serve for arguments dependent on time comparison, and there is no guarantee that the patterns of development, including regional patterns of income inequality and of agricultural production, will follow trends established elsewhere. There is evidence that the general changes in the world relationships of nations consequent upon the development of a few alter the conditions and patterns of development for the rest.

One can, however, see evidence of regional widening of income and agriculture production differences as some progress in improving agricultural technology and in developing the commercial aspects of agricultural production has been achieved. One may trace this partly to an inevitable inequality in technological achievement, which cannot advance equally in all sectors, combined with regional patterning of agricultural enterprises. One may also link it to commercial development and to the fact that marketing depends on a network of transport and exchange whose very construction will produce inequalities in development which will persist the more slowly the construction proceeds. Even with a fully developed system there can never be equal accessibility and costing from all points unless an artificial pricing system overriding accessibility costs is introduced. Hence the tendency is for the broad agricultural regions, more especially those dominated by a single commercial enterprise, such as the coffee growing regions of Latin America and East Africa and the cocoa growing regions of Latin America and West Africa, to consist of a series of subregional concentrations or nodes.

The success of certain tree crops in European markets in the late nineteenth century encouraged agricultural development mainly in those tropical countries which possessed good access to the markets, extensive areas of rainforest, preferably near the coast and preferably unoccupied, and a capacity to develop a satisfactory infrastructure for handling the crop. Those rainforest areas which were nearest the main concentrations of population, amongst whom were farmers willing both to innovate and to migrate, which possessed satisfactory soil conditions, mainly clays, and which were on or near railway or major road routes, were the first to be developed and their migrant farmers were the first to increase their wealth by the new enterprises. Not all Third World countries have followed similar patterns of agricultural development or of development of other forms of production. The enormous variation in social, political and environmental condition has created a wide variation in regional income and development inequalities, including:

1. A remarkably rapid growth of wealth in a few small regions and more especially in urban areas in the oil rich Third World countries, e.g. Libya, Saudi Arabia and Nigeria, leaving regions dependent mainly on traditional agricultural methods far behind. Government policy varies in the oil rich countries with regard to the proportion of the newly acquired wealth to be invested in other sectors of the economy and in other regions. Certain other countries have benefited from mineral wealth, although less spectacularly, particularly in recent years with earnings from iron ore, copper and phosphate exploitation. Often a large part, even most, of this wealth has accrued to metropolitan centres rather than to the area of production.

2. Countries in which agricultural exports have been successfully developed, although often associated with a high degree of specialization, sometimes even a 'one-crop dominated' economy. These may be divided into those countries in which peasant farmers have had the opportunity to develop commercial smallholdings and in which a large part of the wealth earned has accrued to the rural areas, more especially in tropical Africa and Southeast Asia, and those countries with extensive estate or plantation systems, more especially in Latin America, in which much of the wealth earned has accrued to urban-based owners and shareholders. Heavy taxation of smallholders in some African and Southeast Asian countries to provide the funds for urban based development has produced regional contrasts not dissimilar from those in certain Latin American countries. One might further divide these groups of countries by their differences in degree and kind of 'dependence' on more developed countries both politically and economically,

which influences considerably the formation of central places in less developed countries and through them the regional character of agricultural development (Rosciszewski, 1974).

3. Countries with expanding manufacturing sectors such as Argentina and Brazil, which might be regarded as 'more advanced', but which have frequently found the funds for industrial investment from the profits of agriculture. As Gilbert has commented, regions such as northeast Brazil, the sierra of Peru and the coffee growing regions of Colombia are in addition having to subsidize industrial growth by paying the industrial areas higher prices than they would for imported products and are often obliged to sell their foreign earnings at an unfavourable exchange rate laid down by the national government (Gilbert, 1974, p. 230). They also export the stronger elements in their labour force, encouraging productivity decline. Such regions become even poorer than they would have been had industrial development not been attempted, except in so far as they may receive some income from emigrants to the towns and some share of the national expenditure on services. Undoubtedly for some rural areas, even though they may have developed commercial agricultural systems and a high level of productivity, further national development located elsewhere is inclined to pauperize them. A tendency to convergence has been shown in some Latin American countries, notably Brazil and Mexico, with decreasing regional relative disparities, but even for these, as Gilbert has shown, absolute differences are still widening (Gilbert, 1974, pp. 219–35).

4. Countries with huge areas of rural economic depression, mostly associated with massive overcrowding, where regional inequalities will be widened as much by the steadily worsening situation in such areas as by improvement elsewhere. In India the development of an advanced scientific and engineering technology which is associated with a growing urban wealth, is in strong contrast with the primitive farming techniques and worsening poverty of millions of peasant farmers. The Green Revolution has provided a sort of intermediate technology which has brought wealth to some farmers and a few regions, mainly to those areas of larger holdings, more especially those of over 10 hectares. A great mass of petty and fragmented smallholdings has been unable to share in the benefits of the

new techniques and, having reached a point where with current methods intensification can go no further, tends to suffer yield declines as subdivision under population pressure reduces their size. In some cases very low population density discourages development because of the high infra-structure costs in relation to population without massive migration. Interior Tanzania and huge areas of Zaire, which are not otherwise lacking potential for agricultural development, have tended to become stagnant, leaving small population 'islands' with little hope of raising income levels. The problems of interior Brazil are barely touched even by a massive investment in roads and in the development of an interior capital city. The West African 'middle belt' has shared in development only in so far as the transport routes crossing it have provided a few points of advantage or where special movements such as the downhill migration of peoples from plateau areas of restricted agricultural potential have taken place (Gleave, 1966).

It has been claimed that in the Third World the most important decision making level is not local nor regional but national, that the key decision maker is not the entrepreneur but government, and that government services to farmers 'serve as the delivery system through which the necessary inputs or means of production are made available . . .' (FAO, 1974A, p. ix). In most Third World countries there are very few large farms, which elsewhere have been major sources of innovation. Where large farms do exist they have frequently been regarded as profitable only where costs have been kept to a minimum, that is low-input systems with little or no attempt at innovation have been preferred. Little research or conscious attempt to improve productivity has come from the farmers themselves except when, as discussed in Chapter 1, there has been a considerable incentive to change, usually in the promise of extremely high returns to compensate for the estimated risks involved and generally deriving directly or indirectly from overseas trade. It would seem that the great age of rapidly expanding markets for tropical and sub-tropical produce in the more developed countries is over, although rising standards of living and industrial productivity in other countries on the road to becoming 'more developed' could increase demand still further. 'Green Revolutions' have incentives and promise of their own, but generally the long road of raising agricultural productivity, earnings and job satisfaction throughout the Third World involves much more a gradual and widespread attempt to innovate in farming and to change the circumstances in which farming operates than a few sudden and highly localized transformations, whose claims have sometimes been widely exaggerated (see pp. 156–7). Government has become the main agent of such change not only through policies on agricultural production, land and infrastructure, which will be treated separately, but also through the degree of political stability it can achieve, through its policies on regional planning and investment, through sectoral differences in the allocation of resources, and through policy on taxation, revenue and overseas trade.

GOVERNMENT AND AGRICULTURAL TAXATION

The importance of the agricultural sector for taxation in most Third World countries lies above all in its size. For many countries it is overwhelmingly the chief source of revenue. In some countries the agricultural sector is taxed lightly but in others heavy taxation has been introduced, partly to satisfy revenue needs, partly because it has been thought essential for development policy to transfer resources out of agriculture and partly because it was also thought that taxation would act as a spur, encouraging higher productivity and more efficient land use. In *Theory of Economic Growth* (1955, p. 231) Lewis argued, 'If it is desired to accelerate capital formation at a time when profits are still a small proportion of national income there is in practice no other way of doing this than to levy substantially upon agriculture, both because agriculture constitutes 50 to 60 per cent or more of the national income, and also because levying upon other sectors is handicapped by the fact that it is desirable to have these other sectors expand as part of the process of economic growth.' Objection has been made to this thesis and also to the general assumption that there can be universally desirable characteristics of the tax system in the Third World. Increased investment may not lead to growth. Increased savings may lead to underemployment and underutilized capacity (Bird, 1974, pp. 3–6). Taxation may even have inhibited agricultural development because it so frequently bears most heavily on the main area of hitherto successful commercial development, that is on export crop production, in part because of a preference in many countries for indirect methods of taxation—customs and excise revenues, purchase tax, 'trading surpluses' of marketing boards—which are more easily levied upon export crop production than upon any other branch of agriculture. Export crop taxes have become the most significant way of taxing agriculture in those countries with considerable export crop production, despite a general decline in the fiscal importance of export taxes in the Third World (Bird, 1974, p. 33). Generally less than 20 per cent of the revenues of less developed countries have been derived from direct taxes (Prest, 1970). Whilst heavy rates of taxation on the higher incomes in agriculture may be essential to revenue and desirable in the interests of social equity and may in some instances, such as their effect on absentee landlords, hardly interfere with incentives or economic growth (Kaldor, 1970), there can be no doubt that they often adversely affect the few successful agricultural innovators. They raise the 'high pay-off' target level that is essential to assure potential innovators of a sufficient reward to make effort and risk taking worth while. In effect a taxation policy intended to foster a structural transformation from a 'traditional economy' dominated by subsistence to a modern economy dominated by industry and commerce seems likely to succeed in hindering the structural transformation of agriculture itself, since it is not the 'slack' or unutilized potential in agriculture which in practice is most severely hit. There is also evidence that where poor farmers are adversely affected by taxation they are likelier to swell the numbers of urban unemployed, the new 'slack', than to contribute to either agricultural or industrial transformation (Bird, 1974, p. 16). Taxation of subsistence farmers in order to force the pace of development has always been difficult because of the problems of raising cash and of the poor returns for the taxation effort involved. Personal taxes such as poll and hut taxes have been restricted mainly to Africa and have declined in importance; few countries have land taxes, largely because of universally weak valuation systems (Bird, 1974, p. 75), and in those few the use of land taxes has declined under political pressures (Kaldor, 1970, p. 168).

With increased urbanization there is a growing market for local food crop production. Often the dealings in urban markets make it impossible to determine the incomes that result, and production for such markets tends to escape the high levels of taxation which in some countries affect export crops. One can see the possibility of a preference for commercial crop production for local markets as a form of tax avoidance, an effect which unfortunately it is impossible at present to estimate, so that the result in zonation of cropping around major towns must remain unknown. Import taxes almost certainly act as a disincentive to all commercial farmers, but again the effects have not been measured. It has been argued that some measure of taxation encourages production and that a rise in incomes may even reduce the productivity of some peasant farmers because they may satisfy their needs with a smaller output (Kaldor, 1970, p. 167), a form of 'target-working' in agriculture. Moreover, governments may be encouraged to keep the prices of agricultural produce low, especially of foodstuffs, in situations where urban incomes are low, political difficulties considerable and industrial growth poor. Generally, however, in the Third World there is much evidence that farmers respond positively to rising prices, especially if a substantial rise in income from higher output seems likely (Elkan, 1973, pp. 107–10), and it would appear that some part of the success, such as it has been, of the Green Revolution in India has been due to rising food prices. In contrast some part of the slow growth in the production of export commodities in recent years has stemmed not so much from disease, poor weather or other environmental limitations, although these have occurred, as from low prices and the disincentives of agricultural taxation.

Some of these problems of location, spatial expansion and taxation in agriculture are well illustrated in Nigeria. Johnson (1968, pp. 99–102) has claimed that governments in Nigeria have acted in effect as landowners and have taxed export crops so heavily that they have probably extracted most of the economic rent. The result has been virtually no income for the farmers to purchase capital equipment except small items, and a tendency to overinvestment in those few enterprises such as rice and poultry where prices and incomes appear favourable. Nigeria's marketing boards assumed responsibility for export crop producer price stabilization and for the development of related industries. They had, however, considerable powers to accumulate and spend funds earned from trading on development projects and exercised these beyond any original limits set. Marketing board earnings became in effect one of the chief sources of revenue for the development budgets of the regional governments (Helleiner, 1970) as Bauer had suggested as early as 1954. The boards appear to have accumulated far more in 'trading profits' than the government earned in tax revenue in the period 1947–54, a major period of expansion in cocoa, cotton, groundnut, palm oil and kernel production. Huge trading surpluses were accumulated especially in cocoa trading in the Western Region and very little of these was returned to the farmers when prices were low. During the period 1947–54 the percentages of potential producer income which were withheld in the form of marketing board trading surpluses were cocoa 21·6, groundnuts 28·1, palm kernels 19·9, palm oil 8·0 and cotton 30·3. Total withdrawals including export duties and produce purchase tax were cocoa 39·4, groundnuts 40·0, palm kernels 29·2, palm oil 17·0 and cotton 42·3 (Helleiner, 1970, p. 419). These latter figures were reduced to a range of 11·2 to 27·1 per cent in 1954–61, but even that was high for a period of very low prices, especially since taxation did not end there. Even allowing that reported farmers' incomes might be only 36 per cent of estimated incomes so that the effect of direct taxation was likely to be low (Taylor, 1970, pp. 517–20), there were still taxes in the form of duties on imported goods and, for those with low incomes, direct taxation by a poll tax. Many cocoa farmers during their most profitable period could hardly have been paying less than half their true cocoa earnings to the government and probably nearly a third during the following period of reduced incomes. Although some of the marketing board 'profit' was spent on research and extension work by the regional department of agriculture, on chemical sprays and other activities connected with agricultural improvement, the evidence all suggests a serious financial disincentive and a probable greater extent of cocoa cultivation with a more liberal taxation policy. One might be driven to conclude that in so far as Nigerian production was reduced by taxation so productivity in the crops affected was promoted elsewhere. Of course there are other implications, including the effect of greater purchasing power on imports, regional investment, transport and marketing and the world trade in cocoa which might have felt recession sooner had the West African marketing boards acted differently.

GOVERNMENT AND AGRICULTURAL INVESTMENT: REGIONAL PLANNING IN RURAL AREAS, RESETTLEMENT AND CREDIT

In agricultural investment, and more especially in rural regional planning, resettlement and the provision of credit, the effect of government action on enterprise location may be seen more directly and in many cases may be measured, although the overall national effect may be much smaller than that achieved by the manipulation of taxes. There are many other aspects of government investment in agriculture, including research, innovation, advice, education and infrastructure, but these either have peculiar spatial aspects of their own and are dealt with separately below or are better treated in relation to other topics or even at other scales.

In Brazil the considerable regional disparities in income have prompted government spending on regional development programmes intended to reduce the great income range and to promote development, especially in the poorer regions. The problem was complicated by the fact that many forms of economic activity were highly localized, so that regional differences in income were associated with sectoral differences in production (Higgins, 1972). The Brazilian government has pursued a policy of encouraging spread effects by improved communications, especially roads, between the industrial heartland and the North, the Central Plateau and the Northeast (the net effect could of course be a greater backwash movement!). It has created regional development agencies to promote economic development in specific regions, notably in the

São Francisco and Paraíba valleys, the Amazon Region, the Southwest Frontier and the Northeast. It has also supported programmes for agricultural colonization, such as the SUDENE (the Superintendency for the Development of the North-East) settlement in western Maranhão, and invested more in agricultural research, education and extension work.

A general consensus in favour of development emerged in Brazil in the fifties and especially for development in the North-East, which had suffered poverty and famine for over a century (Gilbert, 1974, p. 240). The investment in rural areas, however, is still low compared with urban investment. Brazilian agriculture has served largely as a capital source for industrial development and the rural areas have served as labour sources, receiving very little in return (Schuh, 1970). Regional development has often been unpopular with local politicians, weakening their local power base and encouraging land reform. At the centre it may be seen as a cause of higher taxation (Gilbert, 1974, p. 268). One of the problems of devoting more investment to agriculture seems to be that inequality in regional development is not so much reduced as shifted in its locational pattern as new regions are selected for special treatment. In Thailand political unrest in the North-East Region together with evidence of extreme poverty has encouraged a measure of regional investment by government in construction, power, urban services, transport and irrigation, although unfortunately many of the projects were uneconomic and the irrigation schemes succeeded more by providing off-farm employment than by raising agricultural productivity (Bell, 1969; Dixon, 1975). Several of the projects, including the irrigation schemes, have increased rather than reduced income disparities within the region and more attention has been given recently to land reform, and the provision of rural credit and improved marketing in order to strengthen the spread effect. Even the benefits of cheap credit and lower production and marketing costs, however, accrue more to richer farmers than to poorer. In Tanzania socialist development policies have paid particular attention to rural problems and have encouraged agricultural development as part of the general scheme of promoting *ujamaa* villages or co-operative and communal settlements practising collective farming. The government hopes not only that the *ujamaa* villages will raise levels of agricultural production but will try new crops and expand dairy farming as part of a programme of agricultural diversification. The programme has become highly regionalized, being much less popular in developed agricultural regions, where small individually operated commercial farms have been highly successful, than in the more backward areas looking for government support, that is, in the Mtwara, Lindi, Dodoma and Iringa districts. Some of the villages are oriented to defence and others contain refugees from Moçambique (Connell, 1974). Despite the efforts to promote collectivization most money from the Regional Development Fund has gone to improve productivity in areas where individual farming is dominant, since these are the areas with the highest regional population totals, on the basis of which money has been allocated.

The provision of agricultural credit has become one of the most important of government activities in the promotion of agricultural development. Most Third World countries do not lack sources of rural credit and in some, as in Brazil, which has already been cited, capital accumulated from agricultural activity has been used to promote not only agriculture but service industries and manufacture. The most striking credit programme in Brazil was that of the ACAR (the Rural Association for Credit and Assistance), which by 1964 had reached nearly a third of the farmers in Minas Gerais and which provided a 'package' combining supervised credit and agricultural extension. It was, as Hayami and Ruttan (1971, p. 273) have commented, a model programme but disappointing in performance, for with loans at between a third and a half of normal interest rates, farmers tended to 'overinvest' in capital assets and maximized net worth rather than net productivity. They achieved a lower level of technical efficiency than non-ACAR farmers (Ribeiro and Wharton, 1969). Undoubtedly mistakes have been made concerning the effect of cheap credit and the extent of small farmer savings, but generally the lack of capital has been one of the chief factors inhibiting agricultural development in many areas, more especially where production for subsistence is dominant, and difficulties in credit worthiness combined with high risks and small turn-over have discouraged the banks from lending money and kept local money-lending interest rates high. National credit institutions for farmers, sometimes supported by international aid agencies (pp. 113–16), have frequently been seen as essential for agricultural improvement, usually with the provision that they must be supported by reforms in land tenure, taxation, agricultural extension and marketing, but often without any serious analysis of saving and investment behaviour in rural areas (Hayami and Ruttan, 1971, pp. 269–79; Belshaw, 1959). In many cases farmers have sought credit, not to improve agriculture, but to satisfy social needs, particularly the costs involved in funerals and weddings or to promote non-agricultural developments or, and more usually, to relieve existing indebtedness at higher rates of interest to private lending agencies.

GOVERNMENT'S ROLE IN AGRICULTURAL PRODUCTION FOR EXPORT

Many aspects of the development of export cropping have already been considered, but some, more especially the direct promotion of export cropping by government as a major strategy for economic development, have yet to be discussed. In many countries such promotion is today either modest or of very little importance, as governments have sought to develop other sectors of the economy. In the past, however, the promotion of export cropping was almost universal and created in many countries an agricultural and even a social revolution. In some countries, as in the Philippines, for example, described by Hayami and Ruttan (1971, pp. 253–8) to illustrate problems of resource malallocation, policy has shifted twice and once more shows an interest in export cropping. In the 1950s policy shifted from supporting specialization and an external trade in primary products to promoting manufacture, which grew rapidly, accompanied by a sharp deterioration in the balance of payments position. In 1960–2 decontrol measures were undertaken, the peso was devalued and income flows were redirected towards the primary product sectors. In the 1950s food crop production had increased to satisfy rising home demand, but at the expense of export crop production. In the 1960s the pattern was reversed and export crop production increased at the expense of food production. Not until a new rice production technology was developed in the late 1960s was progress made in both.

Export crop cultivation as an important if not actually the sole element of major change and growth in the economy and as one of the major features of government policy in Third World countries reached its zenith in the late nineteenth and early twentieth century. Its foundations go back to the very beginnings of European contact with the Third World and the foundations of overseas colonialism in the trading and production policies of the nineteenth century. The colonial empires were regarded by the metropolitan powers as their chief sources of many primary products, botanic gardens were established to acclimatize exotic plants, colonial governors tried to promote new forms of production and encouraged investment in plantations. In 1690 the Governor-General of the Dutch East Indies planted coffee seeds in a garden at Batavia. One of the first plants grown was sent to the Botanic Garden in Amsterdam and seedlings from that plant were used to establish coffee cultivation in Surinam in 1719. More seedlings were sent to the West Indies in 1728. In 1816 the Governor of Senegal attempted to change the region he controlled from a trading to a planting colony. American seed cotton was introduced and in 1822 his successor authorized the development of an experimental garden and distributed concessions to settlers. By 1858 the then governor saw himself as engaged in a kind of economic war with the British as major rivals in West African trade and asked for an increase in the budget to help develop new crops. Colonial governments extended their operations much further through departments of agriculture, whose main tasks became to research into and to promote export crop production. Grading systems were introduced at buying stations and competition between producers was controlled. In Kenya, for example, European coffee growers were protected from African competition by administrative discouragement of African coffee planting through the operation of the Coffee Plantation Registration Act of 1918. The subsequent Coffee Rules of 1934 (amended 1949) permitted African coffee cultivation in specified areas (Hailey, 1957, p. 834). In many colonies, however, peasant production of export crops was actively encouraged, even to the extent of using compulsion. In former British colonies definite powers were confined to orders for the planting of food crops where famine was threatened. So-called 'moral pressure' was, however, applied in the East African colonies. In former French colonies the compulsory cultivation of certain export crops was supported by legal sanctions and in the former Belgian Congo a system of compulsory cultivation was sanctioned by the Belgian government in 1917 (Hailey, 1957, pp. 630–5).

Compulsory measures involving crop inspection and enforcement have been used not only to ensure export crop planting, but also adequate attention to cultivation and to the introduction and spread of modern techniques, especially of pest control. Cotton production in particular is highly susceptible to disease and pest attack, and has become the object of legislation designed to enforce 'proper standards' of production in several countries. In the Sudan, for example, it was discovered that pink bollworm larvae survived among cotton seed long enough to infest the following year's crop but could be killed by exposure to the hot summer sun. 'Sunning' of cotton seed was enforced by legislation as early as 1917, and was followed by the enforced burning of cotton crop remains, the fixing of dates for burning, the prohibition of planting old seed and of the planting of indigenous cotton varieties in certain areas, and even the prohibition of okra (*Hibiscus esculentus*) cultivation when it was discovered that okra acted as a bollworm host (Tothill, 1948, pp. 507–13).

Taxation was occasionally regarded as a method of inducing an interest in cash crops, inevitably mainly export crops, but usually export cropping was promoted in order to be able to levy higher taxes or export duties and pay for the costs of colonial government. In Latin America governments were less active in the early development of export cropping than in largely colonial Africa and Southeast Asia. The situation was quite different with land tenure, the availability of land for expansion and the availability of local capital for investment in plantation production. In several Latin American countries there was also a well established tradition of plantation production for export and a network of dealers. In Paraguay, however, an early cultural and economic isolation preventing the development of an export crop economy and reflecting the political control of Buenos Aires, was developed into a policy of almost complete isolation by the dictator Francia (1814–40), who kept the economy largely at subsistence level. Huge tracts of land owned by government were not sold for development until after the Triple Alliance War of 1865–70 (Pincus, 1968, pp. 5–8).

More recently agricultural export production in the Third World has been stimulated by a return to plantation type enterprise or to planned systems of farming in which smallholders are controlled by a central agency. These have been encouraged and organized by governments, but often with the aid of foreign investment. Such developments are frequently agro-industrial and include the provision of processing plant, the inputs to farmers, trials, training schemes and marketing organization. In Dominica groundnut production has been organized around a factory for processing, which because of a subsidy supplies farmers with inputs at only nominal cost (Antonini, 1971). In Ethiopia huge estates and mills produce cotton and sisal (Yeates, 1973), and numerous 'company' schemes elsewhere produce cotton, paper pulp, flowers, fruit and dairy produce, and raise cattle (Freivalds, 1973). In Kenya air freight has been used in combination with modern organization and investment to produce chrysanthemum cuttings and asparagus ferns for the European market.

GOVERNMENT'S ROLE IN AGRICULTURAL PRODUCTION FOR THE INTERNAL MARKET

The notion that governments in Third World countries should take an active part in encouraging agricultural production for home consumption, both food crops and raw material crops to supply local industries, was slow to develop. It was generated partly by mounting evidence in many countries of food shortage and even the occurrence of famine, partly by concern over standards of nutrition, partly by evidence that export crop oriented policies were taking away resources from food crop production and partly by increasing food imports which had to be paid for with overseas earnings required for expenditure on other commodities. There was also concern to eliminate supposedly wasteful methods of production, such as shifting cultivation, and to introduce more modern methods, including conservation techniques, and there was a fear of over-dependence on one or just a few major export crops, coupled with a preference for economic diversification. The foundations of these developments were mostly laid in the 1920s and 1930s, although there were many much earlier food crop introductions and the use of compulsion to ensure that enough food crops were planted to avoid famine.

In the former colonial territories belief in laissez-faire economics did not extend to the problems of peasant agriculture, where officers of the agricultural departments applied a more paternalistic philosophy. Since the 1930s mounting concern with the world food situation, declining food production per capita in many less developed countries and rapid rises in many cases in food imports have encouraged or even forced many Third World governments into playing a very active role and into considerable investment in raising agricultural productivity. In the 1960s the widening productivity gap between the agricultures of the more developed countries and those of the less developed countries was especially marked by the rising outputs per hectare and per worker of Western Europe and North America and the virtual stagnation of countries in South and Southeast Asia such as Sri Lanka and India (Hayami and Ruttan, 1971, pp. 71–4).

In several Latin American countries yields remained very low, but there has generally been some rise in agricultural output per worker or such output has remained for the most part higher than in Southeast Asia. Nutrition problems amongst the Latin American peasantry, although severe enough by the standards of North America, have nevertheless been less serious than in several tropical African and most tropical Asian countries. More recently, however, concern with agricultural production for the internal market has grown with mounting evidence of massive rural poverty and of a threat to agricultural productivity from the extensive migration from the rural areas to the cities. Most effort has been concentrated on land tenure reform and colonization schemes. Elsewhere, especially in Southeast Asia, there is much more interest in improved husbandry techniques and intensification, although land tenure reform is also frequently a major issue. In Malaya, for example, the need for rice imports, evidence of decreases in the size of rice holdings as the population grows, increased fragmentation, the poor returns from rice cultivation compared with those from other crops, especially rubber, coconuts and vegetables, and the fact that new land was not readily available for rice cultivation although 70 per cent of the country remained undeveloped, prompted the government in the First Malaysia Plan 1966–70 to allocate more than a third of agricultural expenditure for irrigation works in order to increase the area available for rice double-cropping (Ho, 1969). Additional sums were allocated for crop and fertilizer subsidies, agricultural credit and research.

In India massive rural poverty and lack of capital often means that national government or foreign help is essential to initiate agricultural progress. An important factor in increasing productivity has been the provision of national government-built canal irrigation and the introduction by state governments of improved seeds, implements, schemes for the wider use of manures and fertilizers and the development of agricultural research. A somewhat unexpected result of introducing higher yielding rice strains in Madras has been a decrease in the area sown, giving fairly constant production, accompanied by increased fallows, retirement of marginal land and a transfer of part of the land to other crops, including sugar cane and cotton. The total foodgrain production has in fact declined as the area not only of rice but of other foodgrains has decreased and been replaced by cash crops. Part of the millet area, for example, has been transferred to groundnuts (Chakravarti, 1971).

Frequently the promotion of home food production has driven governments to formulate not just a policy of encouragement through advisory services, credit provision and marketing control, but of guidance or even control of agricultural production at all stages and the use of regulations and laws backed by police inspection and the threat of imprisonment to ensure 'sound agricultural practice'. Thus in southern Malawi, for example, regulations were enforced to control soil erosion and ensure the destruction of disease ridden crop remains (Morgan, 1953) and in certain locations box ridging was enforced.

Attempts to improve agriculture by control measures of this kind have been widely practised in tropical Africa and many examples are quoted in Lord Hailey's 'African Survey' (1957). They probably achieved their most extreme form in the 'paysannat dirigé' programme in the former Belgian Congo (see p. 158).

In Africa problems of rising food imports and scarce resources, chiefly labour, have encouraged investment in modern techniques in food crop production, especially the use of tractors. Mechanized farms for the production not just of groundnuts, but of food crops such as maize and rice, have appeared in many African countries and apparently reversed the traditional arrangement of large estates for export crop production amidst largely food producing small-scale peasant farms. These are often state farms, another form of government investment in agriculture. Large mechanized food crop farms and small family farms concentrating on export cropping might be a sensible policy where the export crops involve a high degree of specialization and where high returns are accompanied by low risks, thus inducing peasant farmers to devote most of their land to export crop production. Such a situation could arise where the export crops involved hand tillage methods, as, for example, tree crops such as cocoa, coffee and rubber, and the main food crops were grains. It could have occurred in southwestern Nigeria where local labour resources would have been a major constraint on an agricultural production devoted so largely to cocoa, had large quantities of migratory labour from the northern savannas with their short growing season not been available for planting food crops. It has mostly occurred, however, where imported grains have proved cheaper than food grown on family smallholdings or where problems of storage and weather variability have led to high levels of risk with occasional food shortages and even famine.

All too often mechanized production of food grains in Africa has proved more expensive than local hand tillage or imported foodstuffs, but has been supported by a government subsidy from the profits of export cropping, intended to save overseas currency earnings, but without attention to the effect of the tax or duty on export crop production. In Tanzania some of the former groundnut estates have become large-scale mechanized maize producers, and mechanical methods, introduced as early as the 1930s, are also used to cultivate wheat in the Southern Highlands. The 1969–74 Five Year Plan had as one of its aims the achievement of self-sufficiency in wheat. State farms for wheat production have been created, especially in the remoter areas of the southwest. In Kenya the problem of cyclic production especially in maize, exacerbated by the higher productivity of the new hybrids, has led to a system of controlled purchase by the Maize and Produce Marketing Board (Ogada, 1970). In Ghana also former groundnut producing areas dependent on mechanization were converted to foodcrop production. Efforts to achieve large-scale food production by means of state farms have generally not succeeded and have been replaced by complex privately operated farms (Gurnah, 1973), chiefly for the cultivation of maize. Government schemes to introduce or improve rice production have been a feature of agricultural policy in Sierra Leone, Liberia, Guinea, Gambia, Senegal and Mali. In Sierra Leone a national scheme to increase rice production and reduce imports has involved the clearance and drainage of both coastal and inland freshwater swamps, the breeding of improved strains, the development of cultivation techniques to suit local conditions, the introduction of tractors, the creation of a government rice purchase scheme to control prices and the introduction of subsidies to encourage vegetation clearance and the use of fertilizers, tractors, bunding and transplanting. The scheme is also intended to provide a measure of soil conservation by reducing the amount of shifting cultivation in the uplands (see p. 167).

Import substitution has been a major factor in schemes not only to promote rice cultivation, but to achieve increased production of wheat and maize, more especially in Mexico and Pakistan, and in extremely widespread attempts to grow sugar cane in huge estates served by modern mills particularly in tropical Africa (O'Connor, 1975). In Jamaica the extension of food importing to vegetables, stimulated by the growing tourist trade, led to government attempts to promote vegetable production in order to reduce import costs, and was even accompanied by the cultivation of 'winter' vegetables by subsidiaries of United States firms for export (Due and Gehring, 1973).

LAND TENURE AND LAND HOLDING CONCENTRATION

Land tenure is an important question at several levels, particularly when discussing the farms themselves. In the Third World land tenure exists in considerable variety and several of the tenure forms have been considered obstacles to agricultural change and to economic growth. They have become objects of government policy which has frequently seized on land tenure as a key constraint in agricultural development programmes. Hence arises the strong interest in tenure at the national scale, more especially in recent changes in tenure, prospects for further change and their relationship to the size of holding and to agricultural investment and practice. This interest has been reinforced by the fact that in most Third World countries land is a much higher proportion of total wealth than in more developed countries.

Land reform is, however, never a question solely of wealth or of production efficiency. Often it is regarded as important not as a means to agricultural improvement but as an issue of social justice (Seers and Joy, 1971, p. 191). In consequence, many land reform programmes are concerned mainly with social or political issues and can result in the creation of less efficient systems of production. Thus some governments, especially in former colonial tropical Africa, have seen the creation of private rights in land instead of a usufruct as an object of land reform, whilst governments committed to socialist programmes have seen land reform as a device to create agricultural communes or forms of group enterprise. In Latin America land reform often means dealing with the *latifundias* and absentee landlords. In Egypt and India land reform is restricting the power of the landlords, giving land to tenant farmers or restricting the size of holdings. Land reform may involve a variety of activities including land transfer from one group to another; more rational redistribution of land amongst existing holders, e.g. by consolidating fragmented holdings, or redistribution as a means of breaking up large estates; the creation of new land holding units, often associated with resettlement projects; the elimination of multiple ownership; land nationalization and the removal of private landlords; control or removal of land rent. Fragmentation is not always, however, the result of pressure on land resources; often it is the result of the expansion of certain farm businesses due to their success. Nor are all landlords trying to bar progress or doing so in effect by charging exorbitant rents. Many are concerned with improvement and seek to innovate. Opposition to land reform programmes is usually strongest in countries where political power is largely a product of land ownership, particularly in Latin America, where an enormous amount of land reform legislation has had very little result because few acts have been implemented (Grigg, 1970, p. 103). Land reform is often a cheap means of agricultural development because it is concerned with structural changes, or it is a device for persuading a population that government is playing a major role in social and economic change.

Six broad categories of land tenure with distinct relationships to systems of agricultural production may be recognized (World Bank, 1974, pp. 12–16):

1. South Asian and North African estates or landlord-tenant arrangements with various kinds of tenancy, mainly forms of sharecropping in which ownership is confined to an élite minority and productivity both per hectare and per capita is low. Most tenant farmers produce only a small marketable surplus apart from their subsistence and the landlords' share. Labour intensity is high and capital intensity low.

2. Latin American estates comprising both landlord-tenant and owner/manager-labourer structures together with mixtures of these. Again ownership is confined to an élite minority but production is mainly for commercial purposes, although subsistence is still an important feature. Both labour and capital intensity are low.

3. Usufructuary or use-right arrangements whereby land is held mainly by lineage or by a particular social group whose members share the use of the land and cultivate mainly for subsistence. These occur chiefly in tropical Africa and Southeast Asia. Labour intensity varies but capital intensity is low. While being the most egalitarian, they frequently discourage innovation, since this often requires a degree of individual initiative and enterprise. They can hinder commercial develop-

ments, especially tree-cropping, where a given piece of land may revert to the social group at the end of a cultivation season. Often use-right systems are associated with systems of individual tenure whose importance may be considerable amongst communities experiencing land shortage. In many Latin American countries use-right systems are unrecognized, although traditional pre-Columbian agricultural techniques related to sharing land or to a free use of land in areas of extremely low population density still survive. Often traditional cultivators producing mainly subsistence crops are on land which is owned but has never been otherwise exploited and are classed as squatters.

4. Plantation and ranch types, normally owned by foreigners and sometimes by the state, employing wage labour and producing mainly for export markets. These systems are often capital and labour intensive and can be highly productive. The land tenure problem associated with them derives from their frequent occurrence in or near areas of agricultural land shortage where their creation may originally have involved the displacement of peasant farmers. Some land redistribution policies aiming to break up plantations and create peasant smallholdings have resulted in a decline in total production.

State-owned plantations, farms or ranches include holdings taken over from private ownership and newly created holdings, often designed to solve immediate problems of particular commodity shortage or to substitute machines for labour or to achieve economies of scale. In many cases they have been created as part of a general policy of 'socializing' agriculture. In others they are part of a mixed economy in which government agencies share agricultural development with the private sector, including foreign investment. In Iran, for example, the current land reform programme includes the break up of large estates following the distribution of Crown lands to smallholders in the 1950s. It also includes setting up village co-operatives and the creation of large state farm corporations, some of which are joint ventures with foreign capital participation to produce special crops, such as asparagus, for foreign markets, or to satisfy rising domestic demands for rice, tea and cotton. Very large livestock breeding stations have also been created, chiefly for increases in lamb production and dairy produce.

5. Socialist forms of tenure, where land is vested usually in

the state or a very large group, including collective farms and communally organized village holdings of the *ujamaa* (Tanzania) type. These are increasing, especially in Southeast Asia, as new socialist or communist governments are established and institute their programmes of agrarian reform. In some cases the attempt to socialize agriculture and land is linked to an attempt to revive pre-colonial or pre-capitalist traditions, often linked with a belief in the superiority of rural life and morals and a deliberate encouragement of subsistence production in communes as a means of isolation from capitalist influence (see p. 205). Often socialist governments have created state farms or extremely large state-controlled units, the socialist equivalent of the plantation, alongside the collective farms, either for experimental reasons, or, more commonly, in order to produce by modern mechanized methods large quantities of food staples for which a shortage is feared (see pp. 164–6).

Co-operative farming systems can be regarded as the product of socialist policies, and in many cases they are, as, for example, in Sri Lanka, where the government redistributed 200 000 hectares of 'under-utilised' land as group farms organized on co-operative lines (Pickstock, 1974). Often, however, the incentive for co-operative farming lies in the opportunity for greater profit by effecting scale economies. Co-operatives can therefore occur in capitalist systems as a response by small farmers to economic problems and are frequently encouraged by governments whose policies are not aimed at creating a socialist society. Many co-operatives exist mainly as societies for purchasing inputs in bulk and for making more advantageous sale arrangements than small individual farmers can do. Often, therefore, the creation of agricultural co-operatives has not involved changes of land tenure. In some land reforms, however, notably in Egypt, co-operatives have been thought to reduce the disruptions which might have resulted from the removal of landlords' control (Issawi, 1963, p. 166).

6. Private ownership of land, linked to commercial or market-economy agriculture, in which land is held by individuals and can be bought or sold. An enormous range of size of units exists, in large part because successful farmers may increase the size of their business by land purchase and by so doing may not only increase their income but achieve econo-

mies of scale. In many cases, either through pressure on land resources, or through a preference for certain more favourable locations, private ownership is associated with higher levels of intensity of production and with more effective techniques in order to increase output. Economic development associated with increased commercialism has frequently encouraged the extension of the private ownership of land, associated with rising numbers of landless labourers and migration of the rural poor to the towns. The change to private ownership has sometimes increased the efficiency of agricultural production, concentrated it much more into preferred areas, especially around large towns, and created social inequities.

Some of the Latin American countries have or have had the highest concentrations in the world of agricultural land in the hands of a few private owners. Table 3 shows for some of these countries and some Asian countries an index of concentration of landownership and a comparison with holding size, employ-

TABLE 3

CONCENTRATION OF LAND OWNERSHIP, AVERAGE HOLDING SIZE, EMPLOYMENT PER HECTARE AND LANDLESSNESS IN LATIN AMERICA AND SOUTH AND SOUTHEAST ASIA

	Data year	*Index of concentration of land ownership*	*Average holding size (ha)*	*Employment (man-years per ha)*	*Landless workers as percentage of active population in agriculture*
Latin America					
Argentina	1970	·873	270·10	0·01	51
Brazil	1960	·845	79·25	0·05	26
Colombia	1960	·865	22·60	0·10	42
Peru	1961	·947	20·37	0·10	30
Uruguay	1966	·833	208·80	0·01	55
Venezuela	1961	·936	81·24	0·03	33
South and S.E. Asia					
Taiwan	1960–1	·474	1·27	2·05	n.d.
India	1960	·607	6·52	1·22	32
Iran	1960	·624	6·05	0·32	25
Japan	1960	·473	1·18	1·45	n.d.
Pakistan	1960	·607	2·35	0·96	29
Philippines	1960	·580	3·59	1·25	n.d.

Source: World Bank, *Land Reform*, Rural Development Series, Washington, 1974.

ment per hectare and landlessness. Whilst output per farm worker in these Latin American countries is in general above world average, and in the mainly pastoral countries of Uruguay and Argentina, extremely high, yields per hectare are normally low. The land-ownership system is still remarkably similar to that of the colonial period despite large-scale land reforms, especially for example, in Mexico, and despite extensive colonization programmes involving the establishment of small to medium size family farms (Gilbert, 1974, p. 130). In a situation of land abundance extremely high proportions of the rural population are landless and work as labourers or have only limited opportunities as squatters on smallholdings. Currently nearly 50 per cent of the active population in agriculture in Mexico is landless. In Argentina and Uruguay the figures are 51 per cent and 55 per cent, and in Chile an extraordinary 66 per cent. The landownership concentration guarantees large numbers of *minifundistas* and landless labourers, a huge casual labour market willing to work for low wages on large farms and estates or forced to migrate to the towns. The situation encourages the maintenance of low input systems and low yields, and discourages agricultural innovation. In effect in these countries large holdings have been detrimental to agricultural growth (Gilbert, 1974, pp. 132–4).

In Mexico land reform has its roots in the 1910–15 revolution and the 1915 decree of the Carranza Government that villagers had a right to have enough land for their needs even if this had to be expropriated from adjacent large properties as a *dotacion* or outright grant. Village groups or *ejidos* with usufructuary rights were created as a result, and new villages formed as people moved within the 7 km range of large properties required by the Act. Most of these *ejidos* were formed in the 1930s. Since then more smallholdings with privately-owned rights in land have been created and whilst the share of privately-owned land has not increased the number of private proprietors has risen, nearly doubling between 1940 and 1950 (Gilbert, 1974, p. 151). Substantial regional differences in land quality combined with varying success amongst the *ejidatarios* have created inequalities of income although these have been claimed to be less than those amongst private farmers (World Bank, 1974, p. 68). The changes have also led to the growth of a large landless labour force. By 1960 half the landowners were *ejidatarios*, but whereas in the early years the *ejido* movement had increased productivity, by 1960 there was evidence of lower yields and poorer technology on *ejido* than on private farms and a static farming situation (May and McLellan, 1972, pp. 16–19). Barraza Allande (1974) has claimed that the duality in public and private investment in Mexico has led to polarization of the geographical development of agriculture with high capital-labour ratios and yields in the northern Pacific and northern zones and low elsewhere.

In South and Southeast Asia land reform programmes are aimed at a rural poverty which is related in part to land-ownership and rent problems, but in part also to scarcity of agricultural land, the very small size of the holdings and a low level of rural commercial activity. In most southern Asian countries very high proportions of land are in tenancies and even in countries like Thailand, where the proportions of tenancy are low and decreasing (two-thirds of the landholders in 1963 were owners and more than 20 per cent were at least part owners), tenancy is still an important issue in the main rice growing area of the Central Plain, especially near Bangkok, where irrigated land has been bought by urban landlords. Outside the Central Plain tenancy is insignificant because of the government policy of disposing of state lands to cultivators as smallholdings (Sternstein, 1967). In India land reform has been aimed at the abolition of the *zamindari* system (that is the landlord system created by the British from the revenue-collecting organization of the Moghul rulers), the fixing of rents, establishment of tenant's right to purchase, creation of landownership ceilings and the consolidation of fragmented holdings. The *zamindari* system has been abolished but rents are now paid to government, and most of the tenant farmers have found little economic improvement from the change. More than three million tenants had purchased their holdings by 1961, but rent fixing has not been successful and land ownership ceilings have not been strictly enforced. Consolidation has succeeded in 'rationalizing' holdings occupying nearly 28 million hectares and has been claimed to have contributed to a growth in productivity in three northern states (World Bank, 1974, p. 64).

In Pakistan land reform campaigns left many former estate

owners with 200 hectares and compensation money. Some of them invested in machinery and adopted more intensive farming techniques in order to raise the incomes derived from their reduced holdings to a level comparable with their former earnings (Marshall, 1970). The introduction of high yielding varieties of grains must have seemed to many farmers the answer to land loss as a result of tenure reforms, and the accompanying expansion of irrigation techniques undoubtedly helped to foster a labour shortage, already created in part by migration to the towns, and some decrease in the landless due to land reform.

In the Middle East and in North Africa rural poverty and land tenure problems are closely linked. Land tenure issues are also central to the problem of sedentarization (George, 1973 and see pp. 195–6) or more permanent settlements for pastoral nomads in an increasingly commercial and services conscious society. In tropical Africa land tenure problems, more especially holdings consolidation, are important issues in several countries, although the problems of rural poverty and pressure on land resources are normally not as great as those of Southeast Asia. Kenya in particular has achieved striking progress in consolidation and in the creation of a new reformed smallholdings system with generally larger farms claimed to be economically more viable. Land reform was undertaken by the colonial administration in 1954 with the ending of the Mau Mau 'emergency' and, following the general acceptance of a policy of land consolidation, development intensification and planning of holdings with a rapid introduction of commercial crops as outlined in the so-called Swynnerton Plan (Clayton, 1964). After independence in 1963 the programme of reform was expanded, involving the resettlement of African farmers on the larger farms previously owned by Europeans and the diversification of export output. About 400 000 hectares of former European land was divided into smallholdings and 2·8 million hectares were consolidated. Coffee, pyrethrum, maize, wheat and livestock products have all increased in output and a class of prosperous smallholders created, unfortunately together with a rural landless population some 16 per cent of the total and a huge drift of landless people to Nairobi. Legislation requiring employers to find extra jobs for unemployed workers has been only partly successful in reducing the numbers, and many holdings are still very small. In 1973 it was estimated that 25 per cent were less than 1 hectare and 50 per cent were less than 2 hectares, occupying altogether less than 4 per cent of the arable land total (World Bank, 1974, p. 67).

Land reform has not always been attended by the improvement in productivity and living conditions that its advocates have sought. In some cases it has resulted in high densities of farming population which in turn have led to soil deterioration and even erosion. Thus in highland Bolivia land reform effected changes in the use of land which have encouraged severe damage by goats and excessive clearance for firewood. In one area of the Yungas progressive clearance has taken place up the slope, each field less eroded than the one before as the forest edge is approached. The Revolution has brought about a change in landownership without providing the new owners with an awareness of the problems of management or the conservation techniques needed to solve them (Preston, 1969). Watters (1967) wrote of the deterioration following the occupation of new areas in Venezuela by *campesino* migrants unfamiliar with the local environment who persisted in traditional practice developed elsewhere. Land reform schemes initiated in 1959 on subdivided private estates and state lands gave secure tenure to smallholders on blocks of 10·7 hectares, but failed to achieve higher productivity since the new owners perpetuated the practices associated with their subsistence economies in a situation of poor infrastructure and lack of credit. A few farmers located near markets on good arable land were successful. Kirby (1974) concluded that the real success of the Venezuelan Land Reform Institute (IAN) was not in achieving an equitable distribution of farmland but in achieving some stabilization of the rural population which otherwise would have inflated the army of urban unemployed.

GOVERNMENT AND INNOVATION

Third World governments, through the work mainly of their agricultural departments and often with the aid of foreign agencies, have proved remarkable agricultural innovators despite very small expenditures and effort in relation to the number of farmers it was thought desirable to benefit. Their choices of site for research, experimental farms, extension services and investment in rural transport, marketing and service industries have resulted in the creation of new agricultural regions or, at the very least, in the provision of a framework constraining, promoting or guiding agricultural expansion. Their efforts have not been without criticism, not only of the policies pursued and the amount of investment involved, but of bureaucratic inefficiencies, the frequent urban basis of services intended for rural areas and the lack of interaction between client and change agent because the nature of the civil service structure may encourage an agricultural assistant to please his superior more than his client (Watts, 1969 and 1970; Humphrey, 1970).

Whatever the defects, however, government has been forced into the role of innovator and into creating systems for the diffusion of innovations largely through the lack of private agencies to fill the role. In part this reflects social barriers, preventing the emergence of a class of entrepreneurs as in the lineage systems of Africa and the caste system of India and a lack of incentives to innovate, notably in Latin America. In part it also reflects the high risks of investment in peasant societies and the limited commercialism and consequent poor rewards for introducing innovations. Few private entrepreneurs have appeared from amongst the peasantry themselves or from the ranks of the estate managers or large landholders, who have generally been interested in investing capital in more profitable and reliable enterprises, where possible abroad. Most of the private innovators in the past have been 'planters' from overseas, whether managers or owners, overseas industrialists seeking new sources of raw materials, missionaries, often dependent on the produce of their own farms, some colonial administrators, and commercial entrepreneurs who chose to vary their enterprises by investing in land or who were seeking an investment with long-term potential during a period of commercial depression. Thus the beginnings of cocoa cultivation in Ghana came from the planting of cocoa-seeds by the Basel Missionaries in 1857 and following years, the planting of cocoa probably in 1879 by a Ga blacksmith from Christiansborg who was thought to have worked as a labourer on cocoa plantations in Fernando Po, the activities of rubber traders, some of whom bought cocoa land in 1897, the introduction of cocoa seedlings possibly in 1885 by Governor Griffith and the establishment by Griffith of a government botanical garden at Aburi, which became the chief source of seedlings (Hill, 1963, pp. 170–6). More recently the wealthier members of a growing professional class, including, for example, teachers and civil servants, have invested in modern farming.

Agricultural research in Third World countries has been limited by a shortage of adequately trained personnel, by constant changes in policy as new approaches and emphases have been sought in the hope of development panaceas, and by shortage of funds as investment has been directed more to the rural infrastructure, to commerce or to mining and manufacturing industry. Policy has varied between research on major export crops, on new export crops for trade diversification, on food crops and raw materials for industry as import substitution, on agricultural relocation in order to reduce pressure on land resources, on improved husbandry techniques because of fear of environmental damage from existing techniques, more especially from shifting agriculture, and on soil erosion control and other problems of conservation. It has also included work on pests and diseases, on quality of produce, on food crop production in order to eliminate food shortages or to provide the food resources for an expanding industrial population, and on more varied food crops to improve diets and reduce malnutrition.

A great deal of research at universities and other centres of higher education has no immediate practical objective but simply reflects the interests of the research workers, many of whom have trained abroad. Often research workers are more concerned with 'good husbandry' than with profitability or, especially in systems with a large subsistence sector, with what is likely to be acceptable to farmers given their existing constraints. Remarkably little work has been done in most Third World countries on observing and recording current agricultural practice. In most cases production is merely guessed or estimated from samples of incredibly small size. Hardly any systematic surveys exist and in many cases the size and location of the rural population itself are virtually unknown. There is therefore a situation in which some of the most advanced scientific techniques are being applied by a few research workers to problems which either interest them or which they guess are important, but with little notion of the effects or of resultant application should their experiments succeed. This does not mean that there are not very obvious problems which command attention, such as water control and the development of new rice strains in Southeast Asia, or the control of locusts in Africa, but it does mean that it is in many cases difficult to formulate an overall plan and to order priorities in research. The most obvious problems may not in fact be the most serious —and it is often difficult to locate centres suitable for the application of research since the existing extension and research network may be somewhat haphazard in relation to the distribution of at least part of a country's agricultural production.

Census estimation becomes complicated in a situation where the dearth of information makes the task of stratifying sampling extremely difficult if not in actual fact impossible. The Nigerian 1950–1 sample census of agriculture, for example, depended on a regionalization of the sampling which was based on extremely limited knowledge and a 1½ per cent sample of villages, within each of which 5 per cent of the taxpayers were selected, and from whose fields sample plots of one hundredth of a hectare were chosen for crop checking and weighing.

There have been added difficulties in research arising partly from the tropical environment of most Third World countries and partly from the colonial status of most of them until the 1950s. It was estimated that in the British colonies, for example, a research worker did only a quarter the work of an equivalent worker in the United Kingdom through a shorter career, shorter working day, transfers, interruptions by leave and sickness, and difficulties and delays in obtaining materials and equipment (Masefield, 1972, pp. 80–1). Soil surveys were begun and much valuable information collected, but the task for so few workers was so monumental that we still lack much essential descriptive and analytical material.

The main early successes were in the introduction and

development of new crop varieties and the improvement of produce quality. Generally the early developments were with commercial crops; experiments with staple food crops came later and were followed by work on diseases and pests. Veterinary work early concentrated on diseases and, after some attempts to introduce European breeds of livestock into tropical environments, on the cross-breeding of livestock from different tropical countries and, later still, on the development of pastures and the introduction of fodder crops. Research into agricultural economics and management began in some Third World countries in the 1950s, but in many is still hardly known. In Venezuela, for example, by 1965 most of the agricultural research was in biological aspects, with little practical application until the 1960s. A special report sponsored by the National Fund for Agricultural Investigations recommended a more economic orientation for research staffs and the appointment of agricultural economic technicians as an integral part of research organization (Heaton, 1969, p. 157). In Brazil little economic work has been done; the agricultural sector was largely ignored by state and federal governments, who preferred industrial development programmes; production research was largely lacking; there was little attempt to relate work to the needs of rural people; and the majority of research work was concentrated on variety and fertilizer trials with little hope of breakthroughs which might make substantial impacts on agriculture (Schuh, 1970, pp. 227–40).

Important though much of the research achievement has been in the Third World, progress, especially in agricultural techniques and food crop production, has been extremely slow. The biggest impact has come, perhaps not surprisingly, from international developments and resources, which created the Green Revolution in rice and wheat production, admittedly with an enormous number of problems, but which might have been more successful had much more basic data collection on the agricultural and rural social condition of the Third World countries affected been available.

Agricultural extension services may be provided by a distinct body created as part of a ministry of agriculture and are frequently to be found in former British colonial countries, modelled on the former National Agricultural Advisory Service of the United Kingdom. In other countries the extension services are integrated with research or with marketing, supply, education and regulatory services. In Malaysia they are part of a system of services controlled by the producers—the farmers' associations (Axinn, 1972, pp. 10–12). In Sri Lanka they are linked with research only in the Department of Agriculture, whilst other services are provided by the Department of Agrarian Services (Karunatilake, 1971, pp. 104–5). Agricultural extension informs farmers of new techniques and new methods in farming, advises on farm management and helps farmers to choose methods or systems best suited to their situation. The extension services must have reliable and attractive innovations to offer. Clearly in Third World countries the limited nature of agricultural research has served to limit the quality of the extension services. They must also have customers looking for innovations and again the Third World countries are limited, partly by the frequent existence of very large subsistence sectors and partly by the small size of many farms, which makes the business of experiment difficult and also usually increases farmers' aversion to risk taking.

The keenest innovators have been the large well-organized plantations specializing in a high quality export product, the small to medium sized commercial farmers operating a family export crop business, usually with the help of some, mainly seasonal, hired labour, and the small commercial farmer interested in producing food crops, usually for urban markets. The plantations tend to have their own research and information systems. It is the small to medium sized commercial farmers who want and are able to benefit most from extension services. Few mainly subsistent peasant farmers can take advantage of extension advice and in any case little of it is liable to be directed towards them because funds for extension work are usually very limited and agricultural research and information must be concentrated on a limited range of problems and areas. In consequence there is a tendency for the already successful to receive most advice, and if it is good advice to become more successful still. Here the development processes separate 'classes' of farmers and are in effect creating a new and rising 'middle class' of commercial producers specializing in crops for two particular kinds of market. Much of the effort is wanted near major ports or pick-up points en route for ports or near large towns. The small newly educated class in Third World countries generally prefers urban life and even the staff of a ministry of agriculture may regard farming as an activity low on any social scale. It is therefore hardly surprising that the extension network in many Third World countries is largely urban based. Lipton has referred to the urban bias of the policy makers in India, even where these had a farming background and agriculture was to be given a 'top priority', which remained in effect an abstraction (Lipton, 1973). Preston noted that Pucarani, a provincial capital in the Bolivian Altiplano, was the base for the provincial agricultural extension officer who worked only within 15–20 km of Pucarani and mainly in only four or five communities. He appeared to have very little effect on agriculture and much of the extension work in the province was carried out by mobile teams from the capital, La Paz, which was only 60 km away. He did not, however, appear to help town based estate owners more than poor peasants, although in neighbouring Peru and highland Ecuador a standard criticism of extension workers was the lack of help given to poor farmers. Even in this case, however, one sees the factor of accessibility as having an important influence and one is hardly surprised to discover that over 60 per cent of the information leading innovators to adopt came from friends and neighbours (Preston, 1973, pp. 4–5).

A further problem in the Third World countries is the uneven spread of educational opportunity and the existence in most such countries of a variety of communities, often with limited mobility and sometimes suspicious of strangers, more especially from neighbouring communities. Frequently the extension officer is an outsider whose motives are suspect and who may be thought to threaten and in fact may threaten a community by a scheme of change involving land use planning, settlement and registration of title, and soil erosion and drainage control (Bunting, 1970, pp. 766 and 777–81). Something of the problems in understanding the geography of an extension service in a Third World country may be illustrated by data for Brazilian states of the extension budget per rural inhabitant and the number of rural inhabitants per extension technician (Fig. 21) provided by Schuh (1970, pp. 24–51). Before 1948 the method of extending the results of agricultural research to farmers had been by *fomento* organizations, which furnished technical services at no cost to farmers and inputs at below market price. Frequently the farmers learned little or nothing, budget costs were high, and few, mostly the larger, farmers received help. In 1948 the Association for Rural Credit and Extension (ACAR) was created and began work in Minas Gerais in 'central' Brazil, mainly amongst small farmers. Gradually, however, it concentrated on technical assistance more for the larger farmers and credit more for the smaller farmers. A branch organization was

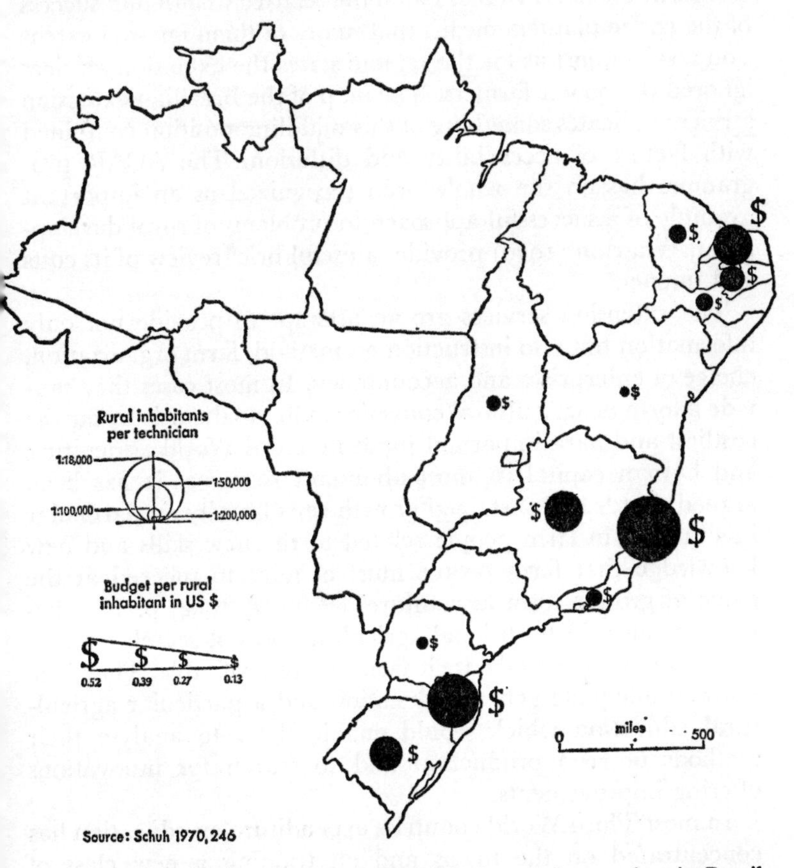

Fig. 21. The distribution of well established extension services in Brazil in 1966

created for the northeast (ANCAR) and in 1956 ABCAR, the national organization, was created. The data mapped are only for the 12 ACAR and ANCAR state agencies which were well established by 1966, although there were in existence six other ABCAR organizations all created since 1960. Note the tendency for three groupings with the Northeast and centre divided by the 'low' of Bahia, a fringe ANCAR state. Note the concentration on coastal Espirito Santo in the centre, with interior Goias as a fringe 'low', and on Santa Catarina in the south. São Paulo and Alagoas have later ABCAR organizations and are not included in the data. In Sao Paulo the relative wealth and success of the coffee planters meant that many of them ignored extension services, just as for the remote states the extension services ignored the poorer farmers. The map of the Brazilian extension services indicates something of this middling position combined with factors of accessibility and diffusion. The ACAR programme has on the whole been recognized as an important example of a successful approach to problems of rural development. Wharton (1970) provides a useful brief review of its costs and impact.

The extension services are an attempt to provide not only information but also instruction on method, farm organization, choice of enterprises and accountancy. In most cases they provide a form of agricultural education. Often labour is by far the costliest and most important input in Third World agriculture and human capital its most abundant resource. It has been argued that 'rapid sustained growth rests heavily on particular investments in farm people related to the new skills and new knowledge that farm people must acquire to succeed at the game of growth from agriculture' (Schultz, 1964, p. 177) because farmers in traditional agriculture do not search for new factors in order to make their farms more profitable. They lack both an adequate general education and a particular agricultural education which would enable them to analyse their methods of crop production and to search for innovations offering improvements.

In most Third World countries expenditure on education has concentrated on the towns and on training a new class of administrators and managers rather than technicians and craftsmen. Most Third World countries have tried to introduce universal free primary education, but on extremely modest budgets and often with little effect in the rural areas. At the secondary school level the urban-rural contrast is usually even greater and even at primary school level the drop-out rates can be startlingly high, again especially in rural areas (Gilbert, 1974, p. 142). The general education offered usually takes conventional forms modelled on imported patterns and thus offers little of value to children who may become farmers. It rarely gives encouragement to develop agricultural skills. In any case many farmers and their children do not want an education which would prepare them for agriculture, but see education as a means to a more renumerative, preferably managerial or professional career in a town. All too often the success of rural education may be measured by the volume of urbanward migration rather than by the rate of agricultural change. In any case lessons from the past such as the successful spread of irrigation techniques in India or of commercial crops in Southeast Asia and tropical Africa suggest that an illiterate peasantry may readily adopt and learn new techniques given sufficient incentives and only a modicum of support by government and private agencies. Today, however, the conditions for agricultural growth have changed. Simple techniques, easily learned, will often no longer suffice. Incentives are likely to be far fewer and a careful appraisal of the problems involved in improvement is needed. Schultz goes further and argues that there is historical evidence of a strong positive relation between the level of skills and knowledge of farm people and their productivity in farming, citing as a Third World example the superior level of productivity of European and Japanese immigrant farmers in South America, and noting that although Third World countries are claimed to lack capital they often seem slow to put new foreign capital to good use (Schultz, 1964, pp. 181 and 186). Schatz has referred to the 'capital shortage illusion' in a study of applications for loans to develop non-agricultural enterprises in Nigeria. His thesis was that there were too few worth while projects, not a lack of capital (Schatz, 1965).

To promote agricultural education and encourage farmers to use more productive ideas and methods, governments in several Third World countries have tried to do much more than offer extension services, and have introduced special training programmes to develop agricultural cadres or 'master farmers'. They have concentrated such activity in selected districts which have been subjected to an intensive package of changes, affecting not only farming but rural education, marketing, credit, transport and farm organization, and have included the building of factories to process farm inputs. An interesting example is the Indian Intensive Agricultural Districts Programme (IADP or Package Programme), which also supported and used the development of agricultural universities, increased use of foreign exchange for the imports needed for agricultural investment and the expansion of irrigation (Malone, 1970). Initially seven districts were chosen, one in each of seven states, totalling over 4 million hectares of arable and tree-crop land, of which over 1·6 million hectares were irrigated. There were 14 000 villages containing nearly 11 million people, of whom 1·3 million were farmers growing mainly rice, but also wheat, barley, maize, millet and sorghum. Each district was provided with an annual agricultural programme containing a package of education and supporting services, including extension, production credit, technical supply and assistance and guidance to farmers, together with an agricultural improvement package of better seed fertilizer, plant protection and water management. In 1962 8 more districts were added, bringing the total to 15 with over 8 million hectares of cropland, 2·9 million farmers and 9000 extension workers (Brown, 1971). In the original 7 districts, as studied by Malone, work began with a promotion programme enrolling some 50–80 per cent of all farmers in chosen villages and by the third year 10 000 of the villages were involved and half a million farmers enrolled. By the sixth year 13 000 villages were involved and 830 000 farmers or approximately two-thirds of the total.

The droughts of 1965–7 hampered progress, the profit gains from the new seeds, especially rice, and fertilizer were not high, but some progress was made. The whole scheme became more attractive after 1967 as farm prices rose and more productive varieties of the major crops were offered (the new High Yielding Varieties of the 'Green Revolution'). In some districts, especially those growing wheat, yields trebled and output rose, accompanied by land use changes as the market was satisfied by smaller areas under a given crop, but in other districts, for example West Godavari, none of the new rice varieties proved both dependable and highly profitable, and the use of new varieties increased only slowly. The IADP districts with their special treatment and virtual priority with regard to obtaining the new inputs and advice became show-pieces of agricultural improvement and an example to other districts, but unfortunately they seem to have had only limited spread as the full benefits of the package programmes have not been offered elsewhere. On the whole the IADP strategy was unsuccessful. Three of the grand total of 15 districts showed overall production increases, but the 15 districts as a group showed no significant growth in either output or yields, not so much through failure of the strategy as such, although this has been criticized as being too supply-based (Desai, 1969), as through failures in the provision of technical assistance for irrigation and in helping farmers with the needed shifts in cropping patterns. The strategy of using scarce funds to pay multipurpose extension workers rather than technically qualified manpower and agricultural research workers may also have been a weakness (Frankel, 1971). The result of national policy has been less the creation of innovative pioneer core districts from which spread effects have extended benefits elsewhere, as an increase in regional diversity, whilst locally the larger farmers are contributing more to and deriving more from the new technology than the smaller farmers.

Programmes like IADP may well be justified in promoting development in only a few areas because of limited resources and the need to avoid risking failure. They have tended to prefer those farmers who have already achieved some progress rather than those most in need since 'the initial effort was to be concentrated in districts where success was most probable . . .' (Malone, 1970). Where they succeed regional or district divergence must result, even if some spread effect is eventually achieved. In some districts, however, the smaller farmers are not far behind the average levels of participation and benefit, so that Allan and Rosing, for example, have been able to conclude from a study in the Bulandshahr District of North West India that whilst the larger farmers do have more advantage and caste is also a factor of some importance, nevertheless there is nothing much to show marked injustice in the development taking place (Allan and Rosing, 1973).

In tropical Africa in the 1950s, mainly in former Belgian Congo and the francophone territories, extension, planning and new inputs were combined in schemes whereby the peasantry were directed by an overseeing authority and forced to adhere to a programme of agricultural improvement or face imprisonment—in the Bambesa region Dumont noted eight days' imprisonment for failure to clear forest to schedule, a fortnight for failure to harvest crops and up to a month for failing to burn cotton plants after picking (Dumont, 1957, p. 50)—the *paysannat dirigé*. In southern Benin Republic, as in many other areas, the *paysannats* have been abandoned, but have been replaced by compulsory co-operative farming systems and by the supervision of agricultural overseers or cadre—the *paysannat encadré* (Dumont, 1966, pp. 55–9). In the Ivory Coast the creation of an agricultural cadre has been linked with army service. Israeli-designed training programmes were introduced by an agreement signed with the Israeli Ministry of Defence in 1961 and based on similar programmes for the Israeli Pioneering Fighting Youth (NAHAL) and Youth Corps (GADNA). The programmes involved agricultural training partly in regional farm camps and partly in 'adopted villages', combined with other 'useful work' and the inculcation of 'proper values' intended to stop the movement of young people to the cities, reduce unemployment and help economic development (Shabtai, 1975). In Western Nigeria in 1960 a farm settlement scheme modelled on the Israeli *moshavim*, or co-operative farm settlement based on individual farms with mutual family help and without hired labour, was introduced. Such settlements were each to contain 50 settlers together with their families. In forest locations they were to grow tree crops on 8–12 hectares and field crops on 2 hectares, and in the savanna mainly field crops on 28 hectares. Tractors were to be used and livestock kept. The settlers were to have two years training at a farm institute and then to work communally and under supervision for two years to establish a settlement. Costs of £4000 per settler were anticipated which it was thought would be repaid at £250 per year. The original farm areas proposed were too large and the settlements too small. In 1962 the revised scheme was for settlements of 100 or more families on smaller holdings. Profits were inadequate for repayment of loans, disputes frequent in the two year establishment period, surveys innaccurate, staff too few and the rate of failure high (Roider, 1970). At best most settlers could hope to earn just £140 per year. Thus the scheme became uneconomic and it was eventually abandoned. In Kenya by contrast the smallholder settlement programme in the former 'White Highlands', which involved supervision and training of farmers, the purchase of land and loans for development, has had some success, although loan repayments have frequently fallen behind schedule and administrative costs have been high. A major advantage here was that the money for land purchase was given by the British Government and part of the development costs came from the International Bank for Reconstruction and Development, the Commonwealth Development Corporation and the West German Government (Maina and MacArthur, 1970). Very active promotion of the production and export of commercial crops, more especially of coffee, fresh fruit and vegetables (especially asparagus, capsicums, French beans and strawberries), tinned pineapple, maize and tea has helped to provide worth while profits. The local textile industry has expanded and increased its demand for cotton. Current development objectives include exports expansion, import substitution, the production of raw materials for processing industries and increase in agricultural employment (McKenzie, 1970).

GOVERNMENT AND LABOUR

Throughout the Third World there are pressing labour problems, sometimes labour shortage, sometimes labour surplus, and very generally labour movement both within a country and across its borders, including both immigration, to work on farms as, for example, of Mossi migrants from Upper Volta to Southern Ivory Coast, or of Spanish labourers, the *golondrinas*, to Argentina, but, more often, emigration from both rural and urban areas of Third World countries to the industrial cities of the more developed world. Labour shortages and migration are common features associating labour with development in the

Third World, although many observers argue that unemployment is the central problem of less developed economies and many more models of a labour surplus situation have been constructed for the Third World than of labour shortage (Lewis, 1954; Fei and Ranis, 1964). However, both the Lewis and the Fei and Ranis models assume full employment in the towns and surplus labour in the rural areas, whereas in many less developed countries exactly the opposite appears to be true (Jolly, de Kadt, Singer and Wilson, 1973, p. 14). A major problem is that no clear definition of unemployment seems possible in the Third World, particularly in the rural areas, and there are few countries whose data are at all reliable. A shortage of job opportunities seems one of the most common features, accompanied by marked seasonal labour surpluses and shortages wherever agriculture is the dominant activity.

In several countries the urban labour force working for government or for industry has been seen to exercise much greater power than the rural work force in obtaining satisfactory wage or income levels (Johnson, 1968; Eicher, 1970). The often spectacular difference in urban and rural incomes has encouraged a migration to the towns, sometimes well in excess of job potential. This migration is frequently directed at the capital city, developing a situation of extraordinary primate growth and allied to major unemployment amongst the educated and semi-skilled rather than the illiterate and unskilled. In Abidjan, capital of the Ivory Coast, 44 000 unemployed were recorded in 1969, of whom more than three-quarters were looking for their first job and were reluctant to accept unskilled work and especially reluctant to return to farming (Roussel, 1971).

Examples of Third World employment situations are:

1. underemployment dominant amongst secondary school higher education leavers, with only small differences between urban and rural areas, as in Sri Lanka (International Labour Office, 1971), and with larger differences in Nigeria, most of Latin America, Pakistan, the Philippines and India;

2. seasonal underemployment or unemployment in rural areas, as in India, Indonesia and Egypt;

3. rural unemployment or underemployment associated with very small holdings or total lack of land as in some Latin

American countries, reflecting an inequitable land tenure system, or in parts of India and Pakistan or in Kenya, reflecting rural overcrowding;

4. poor job opportunities and general underemployment arising from a low level of economic development, limited market opportunities and lack of capital, knowledge and means to stimulate employment and increase productivity, as in several African countries and in many areas in Southeast Asia and Latin America, with little or no transport available.

Few Third World governments have failed to formulate labour policies, although their understanding of their labour situation and problems and the efficiency with which they apply their remedies vary greatly. For agriculture such policies have had major direct effects in increasing or decreasing the supply of labour and important indirect effects through the changing attitudes of farmers to crop production as government labour policy has changed, or through the deliberate manipulation of agricultural production and marketing by government in order to solve labour problems. In Egypt, for example, Hansen (1969), in an important study of employment and wages, showed that the assumption that surplus labour in agriculture amounted to about 25 per cent of the labour force was ill-founded, partly because the women and children included in the estimates for the very active role they play in farming could not represent a reserve of labour available for industrialization, partly because even in some of the most crowded areas and on very small farms men were very fully occupied, with hours averaging by farm size groups 2062–2420 per year (about 7½–8½ hours per day for 275 days a year), and partly because of the heavy demands of the peak labour periods of May–June for the wheat and maize harvest, and September for the cotton harvest. Mabro's general conclusions concerning agricultural wages and employment in Egypt have been cited above (p. 41). He also argued that employers may prefer short contracts since peak labour demand was reached for short periods only and it was common in some cases to pay agreed lump sums for specific tasks such as the harvesting of wheat and cotton. Similar arrangements are made in groundnut farming in Senegal and in coffee and cocoa production in the Ivory Coast, Ghana and Nigeria. Sometimes labour is hired by cocoa farmers not for cocoa cultivation but for food farming operations at times of labour competition between food crops and cocoa, with the advantage that much of the hired labour is already used to food crop farming and not to cocoa cultivation. In a situation of rapid commercial crop expansion, where hand labour is still cheaper than machines, the existence of a source of cheap casual labour able to work for certain brief periods plays a vital role in such expansion and provides that labour with extra income.

The contrast between West Africa and the rest of Africa in the former's juxtaposition of climatic zones, making possible short distance labour migration from areas of only 4 to 6 months rainy season to areas of 9 to 12 months rains, is of vital importance in appreciating the contrast in the development of commercial tree crops between the two regions. In a period of stagnation or limited expansion the importance of immigrant labour is apt to decline, more especially if locally there is a population increase and search for jobs by a landless or nearly landless class. So long as migrant labour is available disguised unemployment is unlikely to occur (Mabro, 1971), and the kind of crop specialization that is normally involved in commercialization can occur. A more 'balanced' labour use, usually involving considerable crop or land use variety, is essential to keep a permanent labour force fully occupied. It therefore follows that attempts to stabilize rural populations, perhaps from fear of unusually rapid urban growth, or to exclude foreign agricultural workers because of unemployment at home or because of political fears, may have unfortunate results in the future if an expansion of tree-cropping or of specialized field-cropping is required. On larger estates some of the advantages both of crop specialization and of combining several crops, together with a more efficient use of labour, may be realized if required by a labour stabilization policy.

It might appear that if migrant work forces decline there will be an incentive to create such estates. In addition to the hired labour problem there is also the question of female and child labour which often works for little more than subsistence on a family farm and may still be cheap relative to hired labour, even when inefficiently used. In this connection land reform, which involves a distribution of land to the peasantry, may not be as inefficient as it may seem in terms of costs if it involves increased use of such labour, but it may well worsen the plight of the landless labourer. Land reform may increase disguised unemployment, especially where agriculture is becoming more commercial, and also increase open unemployment (Mabro, 1971).

In Southeast Asia huge migrations of labour have taken place, often to work on plantations or smaller holdings and often on a long-term or even permanent basis. At times of economic depression such large immigrant minorities have provoked unrest, especially where, as with the Chinese in Malaya and Indonesia, many of the immigrants have become successful in trade, commerce and moneylending and have begun to occupy powerful possition both economically and politically as middlemen. Faced with an ethnic labour problem the Malayan Government embarked on a policy of helping Malays to compete with the Chinese and occupy comparable economic positions, beginning in 1956 with the creation of the Federal Land Development Authority (FLDA), which developed land settlement schemes mainly for Malays.

In Cuba (recently 'centrally planned', but here still classified as Third World) the biggest problem of post-revolutionary development was the acute shortage of labour (Pollitt, 1971) as development schemes created more jobs than the increasing population could fill. From 1958 to 1970 it was estimated that whilst the population increased by 580 000 the number of new employment opportunities rose by 1·2 million. It had been calculated in 1957 that some 36 per cent of the total workforce of 2·2 million was either totally or partially unemployed or engaged as unpaid family labour. The sugar industry was particularly susceptible to marked seasonal and cyclic fluctuations in both income and employment. Fear of sugar dominance together with acute problems of overseas trading relationships following the Revolution led to a policy of agricultural diversification and land reform intended to lead to higher levels of rural employment, greater security for farmers and labourers and increased productivity and incomes. The old system had provided only low levels of both income and employment—'the coexistence of idle land and idle labour' (Pollitt, 1971) restricted output with low intensity of production related to cheap land and a policy of cost minimization. The result of the new policy was reduced agricultural production and exports, with rising imports as change of policy demanded new inputs in the effort to achieve economic diversification. At first the workforce in agriculture rose, but after 1960 it fell, as did the workforce in industry and mining. Labour was divided mainly into services (which rose from 558 000 in 1958–9 to 833 000 in 1964 and included the army of bureaucrats), the social services and the military. General rural 'semi-employment' in 1957 had suggested huge numbers of unemployed, but many of these were the reserve for peak labour demand and had been absorbed in new 'service' occupations, with the result that agricultural labour was short and productivity fell. As Pollitt commented: 'when Fidel Castro marched into Havana in 1959, quite a few economists of repute were publishing development models in which the seasonality of labour did not exist but "surplus labour" always did . . .'

One answer to falling production and a labour shortage was to mechanize. Between 1957 and 1970 the number of tractors increased four times to about 50 000 and were especially important on huge state farms. Unfortunately mechanization had only limited application for the cultivation of sugar cane and was of very little use with several other crops such as coffee and tree fruits. Mostly it resulted in vast increases in the demand for hand labour especially at harvest time, and the harvest of some crops had to be abandoned or undertaken outside the optimum period.

Mechanization is of special interest as a major innovation intended directly to substitute for labour and frequently of major interest to many Third World governments as a technique for modernization, for the clearance of huge areas of land and for the release of rural labour to manufacturing industry. There is, however, a dilemma in that fear of rapidly increasing rural populations combined with urbanward migration and unemployment has also encouraged policies for absorbing more labour into agriculture. Whilst modernization is highly desired a modernization which is labour saving may be seen as only exacerbating current labour problems. As Singh and Day comment '. . . mechanization in a labour-surplus environment is something of a paradox . . .' but it may become necessary through seasonal labour scarcity, even though labour is abundant for the rest of the year, or by the need for timeliness in completing agricultural tasks (Singh and Day, 1975).

Abercombie (1975) has pointed out the accelerating growth in the labour force of the Third World and the tendency of the agricultural labour total to rise, despite a predicted falling proportion of agricultural labour in the labour force through urbanward migration and the growth of manufactures, mining and services. The turning point in the Third World as a whole, where mechanization can make a useful contribution, can come only when the agricultural sector is reduced to 40 per cent of the total labour force (predicted as the year 2006). Schemes such as the Office du Niger Scheme in Mali for irrigated rice and cotton production using tractors for ridging, or the Mwea Tebere Project in Kenya, where rotavators are used, still have a heavy demand for labour and use machinery to extend the area cleared or tilled and improve the timeliness of operations. The latter is especially important in short growing season areas such as in Syria where mechanization has allowed an expansion of the arable area. Such schemes can work when the use of machinery relieves labour bottlenecks, but often, as in several of the ill-fated African mechanized groundnut cultivation schemes, the areas cleared and tilled by tractors have been too extensive for the labour force either to weed or to harvest, and in effect long periods of idleness both for machines and for labour have only increased costs and caused heavy financial loss (Baldwin, 1957).

A second dilemma is that the need for large farm holdings if mechanization is to succeed conflicts with the prevalence of smallholdings in Third World countries and the preference in many for the break-up of large estates with the expansion of the smallholdings area. Attempts to marry mechanization to smallholdings usually take the form either of introducing very small power units such as rotavators and two-wheeled tractors, or the formation of co-operatives combining mechanized joint operation for some farm tasks with individual family operation on separate holdings for others. Except in rice farming in Malaysia and Thailand and in heavily subsidized rice schemes in certain African countries, both small power units and co-operative large tractor and family farming methods have generally proved too expensive, often because they were too sophisticated for peasant farmers and because the necessary servicing, engineering and training skills were poorly provided (Chancellor, 1970; Gordon, 1971; Khan, 1972; Burley, 1974). In Nigeria, for example, the small single axle tractors have proved tiring and costly to operate. They succeeded on small-holdings in Japan, but in totally different ecological and socio-economic conditions (Rana, 1971). In the Philippines hand tractors saved time in tillage and speeded up the cultivation operations for high yielding varieties of rice, making an extra crop possible, but power costs were high and despite low wage labour costs the financial results were frequently poor unless subsidies were provided (Burley, 1974). Despite these problems most Third World governments have attempted to encourage mechanization, and many 'large farmers' prefer using machines to organizing large labour gangs.

One final effect of government policy on agricultural labour in Third World countries is the withdrawal of child labour consequent on the expansion of education programmes. The effect is particularly marked in those areas near to major cities, with well-developed road networks and often the most 'progressive' in the sense of developing modern commercial production and producing crops for export. The loss is often small per farm but not insignificant, as such labour is virtually free since it is usually paid little or nothing and must be provided for in the family budget whether it assists on the farm or not. Cheap family labour is the strength of peasant farming in process of becoming commercialized, more especially where its product must compete with that of well-organized, efficiently producing, large-scale holdings such as plantations. Very young child labour traditionally engaged in special functions such as bird scaring, for which adult labour cannot be spared. There is evidence in northern West Africa, for example, of increasingly heavy losses of grain to *quelea* birds through lack of bird scarers, and of a growing preference for heavily awned although lower yielding varieties of pennisetum or pearl millet.

NATIONAL CONSERVATION

Government policy for agricultural development has frequently been concerned with more than the control of inputs, information and advice or with the attainment of immediate increases in productivity or profitability. Several governments have seen husbandry and its relationship to environment as in need of some sort of control or guidance at the national level or have regarded the development of conservation policies as essential to maintain productivity levels in the long term. In Sierra Leone as early as 1927 (Mackie, Dawe and Loxley, 1927) it was decided that the upland production of rice by shifting cultivation was wasteful of vegetal and soil resources. Expansion of the cultivation of swamp rice was encouraged in order to reduce the demands on what were thought to be fragile upland soils and on traditional agriculture. It was hoped that swamp rice would prove more productive than upland, encouraging a diversion of labour and resources to the valleys and coastlands.* By 1970 it was estimated that only 44 per cent of rice consumption came from the uplands, whilst over 45 per cent came from lowland swamps, mostly freshwater, leaving an amount varying between 5 and 10 per cent each year to be supplied by imports. Swamp rice farming can be made to pay, especially if heavily subsidized in order to encourage its extension and to support the use of fertilizers and tractors, but the capital investment and financial risks are high and it is the subsidies that maintain the spread of swamp rice farming. Labour inputs are higher than on upland farms, which normally can manage without any hired labour, although labour is frequently hired, perhaps partly for prestige (Mackie, Dawe and Loxley, 1927; Waldock, Capstick and Browning, 1951; Jordan, 1954; Jack, 1958; Karr, Njoku and Kallon, 1972). Other forms of control of the environment were tried, some of which were widespread in their use in tropical Africa, particularly attempts to eliminate or limit bush burning, usually by fixing the dates when burning may take place so as to favour shrub regrowth. Normally this has meant encouraging early burning, but often there is a

* By contrast in Indonesia the Schophuys Scheme to resettle farmers from overcrowded areas on newly reclaimed swamp lands in southern Borneo made little progress. Schemes which are pre-occupied with *sawah* cultivation have been criticized as of limited application. There are advocates of dry-rice farming with mechanical equipment as a means of developing 'new' lands (Fisher, 1964, pp. 339–41).

liability to damage to standing crops, especially cotton, and late burning may be more effective in relation to the needs of cultivation. The problem has been that burning is undertaken not just for agricultural purposes, but for the control of pasture grass and for hunting. Optimal dates for these purposes rarely coincide. An alternative method has been to promote mixed farming with the idea of achieving an equilibrium state in the use of soil nutrients or, where the possibilities of livestock keeping seemed limited, to promote techniques such as green manuring. Most of these attempts have not been particularly successful, mainly because of their demands on labour or land and the difficulties of fitting them into existing practice. In the drier savanna lands of West and Central Africa a combination of cattle keeping, chiefly for draught, and crop raising has had some modest success, but generally the feed area required to keep even the few cattle needed for draught or for the supply of milk for the family has to include a considerable area of rough grazings. There are considerable problems during the dry season when grazing on floodland pastures, tree lopping for leaves, the storing of feed or migration elsewhere must be undertaken. Rarely has it been possible to provide enough manure to eliminate the use of self-sown fallows. The most persistent and probably the most successful attempts to promote mixed farming in the African savannas have been in Northern Nigeria, where work began about 1924 and has since been associated not only with a general scheme but also with local resettlement schemes (Morgan and Pugh, 1969, pp. 511–14). By 1964 it was estimated that some 36 000 farmers had taken up the practice. Thus an image of a 'balanced' husbandry without fallows, developed in Western Europe mainly in the late eighteenth and nineteenth centuries, was exported to the Third World mainly through the medium of colonial empire and partly by the use of enforcing legislation.

In part conservation has been practised by taking land out of agricultural production altogether. This has not been popular, especially in countries with increasing pressure on land, and despite the developments of the Green Revolution generally the greatest increases in agricultural production have come from increases in area under crops rather than increases in intensity (pp. 48–52 above). In Nigeria the Forestry Department was created originally to preserve timber and to improve yields from rubber vines. It instituted a policy of reserves, the first in Southern Nigeria being constituted in 1899, with the aim of eventually extending these to occupy 25 per cent of the total area. In addition it tried to develop village wood lots as an alternative to indiscriminate timber cutting and a policy of control of bush burning and of the use of soils. Less than 8 per cent of Nigeria was eventually put into forest reserves, but elsewhere in West Africa higher proportions were achieved, including nearly 9 per cent in Senegal and 14 per cent in Benin Republic (Morgan and Pugh, 1969, pp. 502–4).

In Indonesia the former Dutch administration created a forestry service to save the teak forests from destruction, to save water supplies by preserving forests on watersheds, to plant exotics including species of Eucalyptus, to increase the area planted by a type of *taungya* system (p. 63), whereby smallholders interplanted food crops and trees, and to create forest reserves wherever deforestation threatened to become dangerous. In former French Indochina the colonial government created extensive forest reserves in order to preserve good timber from the destructive methods of shifting cultivation, to save some tree species from extinction and to reduce forest fires. A Belgian answer to the problems of controlling shifting cultivation was to create fast regenerating cultivation 'corridors' with regular rotations and lengths of fallow (see p. 231).

The recognition of the importance of soil conservation in most Third World countries dates chiefly from the 1930s and was influenced by the growing realization of severe soil erosion problems in several areas, especially in the tropics, the increasing influence of Russian soil science, and pioneer work by Sir John Harrison on laterite soils and soil profiles, by Martin and Doyne on silica/alumina ratios in lateritic soils, and by Milne's soil mapping and catena concept (Masefield, 1972, p. 83). Action to control erosion dates from this period, together with attempts to limit activities thought to cause or encourage erosion, such as cultivation on steep slopes and bush-burning, and attempts to foster protective measures, including careful control of cultivation on watersheds and increased use of mulches and manures.

In Pakistan the problems of soil conservation created by

irrigation are particularly severe. The higher water table in the *doabs*, or areas between two rivers, water-logging and salinity led to fears in the early 1960s that much agricultural land would be rendered useless, and prompted government attention to drainage and soil problems and eventually the creation of a public tube well programme in what were defined as Salinity Control and Reclamation Project (SCARP) areas. This involved the installation by government of public tube wells to reclaim waterlogged and saline soil recently gone out of production in the canal-commanded area (Nulty, 1972, pp. 22–3 and 56).

In the Latin American countries much less has been done than in the Middle East, Africa or southern Asia, mainly because most of them have a relative abundance of land, and partly because in several the traditions of estate systems with low inputs, of exploitation of fresh areas as the yields on existing holdings decline and of absentee ownership are still strong. Schuh (1970, pp. 337–9) has even suggested that in Brazil the abundant supply of land has been a factor inhibiting agricultural progress, and has pointed to the general lack of knowledge of both agricultural practice and environment, even in major problem regions such as the Northeast. Venezuela is more advanced in the evaluation of its natural resources than most Latin American countries (Heaton, 1969, pp. 168–73), with reconnaisance resource evaluation studies completed for about 40 per cent of the land area and even a general map of the great soil groups of Venezuela. The essential services required to use this information, however, and to add to it, have received inadequate support.

Despite the general tendency of most government policy in Third World countries towards conservation, in practice there is widespread evidence of continued destruction of vegetation and soil resources. The rainforest, for example, is everywhere in retreat. In South and Southeast Asia the annual loss is over 15 million hectares and in Africa the rainforest area appears to have shrunk by 25 to 30 per cent between 1930 and 1970. Often government policy is to blame. Whilst conservation schemes may be attempted in one location, massive clearance schemes to solve resettlement problems or to provide new highways are being undertaken in others. In the wake of new highway developments, such as the 'Transamazonica' in Brazil, slash-and-burn agriculture, the technique of the pioneer fringe, can bring widespread damage (Manshard, 1975). Formerly sparsely settled Goias is today one of the fastest developing states in Brazil, where settlement is promoted as a means of legitimizing Brazil's territorial claims and despite poor returns (the per capita income in Goias is only half the national average). The Belem–Brasilia highway alone appears to have been responsible for the settlement of between 160 000 and 320 000 pioneers (Katzman, 1975).

4

Agricultural regions in the Third World

In the more developed countries agricultural regions are mostly seen as areas of specialization in the production of a particular farm enterprise or group of enterprises; spectacularly so in the United States, where, as early as the 1920s, huge areas of major crop or livestock production concentration were generally recognized and more formally defined by O. E. Baker (1926–32). Agricultural regions are not discrete objects totally integrating all the phenomena lying within their bounds, nor do they necessarily possess a very high degree of homogeneity of some combination of phenomena, nor are they the inevitable product of the operation of 'natural laws'. They are a convenient way of dividing and classifying space with respect to certain phenomena in order to create a simplified picture of the spatial patterning of agricultural activity (Morgan and Munton, 1971, pp. 126–8). They are essentially an intellectual concept (Whittlesey, 1954) or the product of imagination, but based on real facts of spatial concentration interpreted by partial and sometimes indirect measures of varying degrees of accuracy. Frequently they do contain distinctive landscape features, more especially distinctive patterns of land use, field systems and settlement, but there are few instances of even approximate exclusive co-variation of three or more phenomena of agricultural significance. Significantly, the regions include many features displayed elsewhere and their key criteria are often represented in other locations. Few of them are the product of some centrally-directed plan; most have resulted from the similar decisions of large numbers of farmers living near to one another. We may therefore see many regions as the result not just of particular combinations of factors for which a certain enterprise or particular form of agricultural activity provided a preferred or supposedly optimal farming choice, but as the result of a particular spread or diffusion of information, advice and example. Agricultural regions may be seen to increase or decrease in size, to change their degree of concentration of a particular enterprise or even apparently to move across country. They may also change in shape, split into a number of discrete areas or fuse a number of separate areas together. Although we cannot for the most part identify a particular decision-maker such as a farmer or a government minister with them, as we can with the farm or the nation, nevertheless these are important areal units for consideration in any agricultural study, for they are the product of particular patterns of decision-making, of flows of information and of other spatial systems such as networks of social relationships, of transport and of marketing. They are the spatial expression of organized agricultural activity.

Agricultural regions may be defined by a wide variety of criteria such as agricultural enterprise or enterprise association data, or by measurements of land use, farming inputs, crop areas or quantities, livestock numbers or farming types. In the Third World suitable data are available for few of these and, even for these few, many countries lack detailed information and provide only estimates at some provincial or even national level. Our sources of information and the kinds of regional picture we can make are determined largely by the availability of data which, like the economies of the Third World, are still at a low level of output and quality. There are data for some farming type definitions, but the tendency to classify the great mass of peasant agriculture by its water supply or by the proportions of fallow and crop land generally reflects a situation of marked husbandry contrast with the temperate world and of a lack of farm level studies.

The measurement of inputs and production together with the study of farmer's choice of enterprises has barely begun. All that is available is sampling, often in a largely unknown population and usually with problems of accessibility and of gaining the co-operation of farmers which frequently orient the sampling towards the more progressive farmers and towards those farmers within the orbit of urban influence. For this study the agricultural regions examined will be defined either by single enterprises or by distinctive combinations of enterprises which are held to be dominant; that is, regions which are clearly of greatest importance as a local food resource or as an income earner, but without any attempt to define or measure the degree of dominance, which in most cases is impossible to assess. A great mass of 'in-between' situations will be ignored, even though they make up a very large part of the Third World agricultural total and even though they are of special interest in the study of enterprise choice and diffusion. For the present the author has to accept the information bias in the subject and above all the tendency for better material provision from the more commercial sector. Crop specialization is widespread, however, despite the tendency on many peasant farms, especially those which are largely oriented to subsistence and dependent on rainfall for their moisture supply, to combine many enterprises on a single holding. There are few on which one or two staple food crops, although normally including several varieties, or one or two commercial crops do not occupy more than half the area of the holding. The mixture in a mixed cultivation is largely of secondary intercrops or of a variety of vegetables and fruits in a kind of kitchen garden cultivation. Moreover, as farms in some areas have expanded in size, specialization has increased, not just because expansion is normally associated with commercialization, but because greater distances to additional fields require fewer journeys with specialized cropping than with mixed cropping (see pp. 236–9).

Just as in the more developed world regional patterning of agricultural enterprises or regional specialization is most strongly developed amongst those commercial crops with a strong pull towards markets, more especially to a specialized and sometimes localized network of buyers, so in the less developed countries the most marked regionalization occurs amongst the export crops. It occurs especially amongst those export crops which have proved difficult to incorporate in existing systems of farming or for which for other reasons new systems of farming and land tenure have been developed. All export crops, however, exhibit some kind of regionalization, as do other crops produced for the internal market, whose patterning often tends to reflect the pattern of emerging demand areas, i.e. the towns. Regionalization, whilst most strongly influenced by commercialization, is nevertheless not exclusively its product. Other factors also tend to produce distribution patterns amongst agricultural enterprises, patterns which are often more vaguely defined and possess a somewhat different character, but which are none the less significant features of the agricultural landscape. In these cases the geographical distributions are not organized, as in the regions developed by commercial process, but represent the common response of large numbers of farmers to environment, scope of farming organization and activity, and the influences, even the demands, of the cultures to which they belong. Regional distribution patterns of this kind are like the aggregations which are to be seen in the animal world, where the grouping is the product of common habit or of the need for joint protection, and the population units, in this case the farmers and their families, are one species producing crops for their own consumption, with some small surplus for local exchange, but without any specialization of production or exchange function. These are regions dominated by crops intended almost entirely for subsistence, particularly the major staple food crops, which reflect in part the environmental needs of farming and of the farmers themselves, and in part the geographical pattern of the social organization to which they belong. They have pattern but in a strict sense lack structure, for the pattern is simply a reflection of other geographical distributions and in so far as these are stable so the subsistence crop regions remain stable. The market is fixed, known in advance by the farmers and possesses virtually no costs of transport and distribution. The result is a regional pattern which, although frequently less pronounced or less specialized than that of commercial crop regions, is nevertheless very slow to change, for change must involve the creation of an entirely new regional organization and the reorientation of farming towards different objectives.

Regions of livestock production have so far been excepted from this discussion, whereas generally in the more developed, and mostly temperate climate countries, they would have been included. The separation of livestock farming from crop farming, however, which is largely characteristic not only of the tropical but also of the subtropical portions of the Third World, has resulted in patterns of regionalization which are peculiar to livestock and which involve in the Third World regions of cyclic movement in response to the varying seasonal character of the grazing lands, the search for water and the problems of disease avoidance. Only rarely in the Third World has livestock production involved sophisticated techniques of range management or the development of permanent pastures. Mostly, whether the systems are highly commercial or largely subsistence, inputs have been kept as low as possible with little attempt so far at the improvement of either breeds or pastures. The systems use mainly self-sown grasses and herbs or the leaves of trees, and production fluctuates with the status of the environment. Because of strict separation of livestock production and crop farming in many Third World countries, livestock regions have frequently proved easy to define in general terms, although occasionally difficult in particular terms of the kinds of livestock produced, as some pastoralists have kept more than one kind of animal. In most areas, however, even subsistence pastoralists have tended to specialize in livestock production if only because the pace of movement, the water needs and the kinds of herbage preferred by, say, cattle as opposed to sheep and goats are different. There is an additional problem: in the Third World data for livestock production are even more deficient than those for crop production and even where available usually refer only to one time in a yearly cycle of movement.

AN ANALYSIS OF AGRICULTURAL REGIONS: TYPES OF GEOGRAPHICAL DISTRIBUTION PATTERN

It could be demonstrated that every crop has its own pattern of distribution and that for each crop the pattern is location specific, i.e. each location has its own distinctive supply of inputs, organization of demand and constraints on production. Nevertheless, despite these distinctions and despite the complexity of the overlaps in crop distributions and the variations in joint-enterprise farming systems, there are similarities in certain forms of geographical distribution and regional organization. These are sufficient, despite the complexity, to permit generalization about the appearance of agricultural regions in the Third World and their structure. Much of this generalization is empirically based, that is it rests on a common experience in several locations, but some of it is derived from observation of production and marketing techniques, which make possible conclusions with regard to crop locations derived from an understanding of the relationships involved.

Subsistence and local market crop regions

In their most extreme form such regions, as discussed above, lack structure and reflect mainly the distribution of population and the environmental constraints of the crops concerned, for each cultivating household produces the dominant staple food crop entirely or almost entirely for its own consumption. One may see the pattern clearly in an example such as the distribution of floodland rice in India, where it is constrained by the availability of surface and sub-surface water supply and within that constraint reflects the pattern of population distribution. The strongest concentration of both rice and population is in the Ganges Valley, but important smaller concentrations are in the valley of the Mahanadi and more especially in the delta lowlands of the Mahanadi, Godavari, Krishna and Cauvery. Minor concentrations occur on the narrow western coastal plain and in the upper valleys of the Deccan. These concentrations are almost all identifiable with areas of floodland where water control may be achieved and where high yields make possible very small holdings, usually rather more than two hectares in area, on which farming families may totally depend for their food supply. The correlation, however, between the two distributions, population and rice, is limited. Alternative foodgrains which may be grown on floodland include wheat, maize and sorghum, and the importance of these in relation to rice varies not only according to taste preferences or environmental suitability, but in relation to the tendency to double cropping, more especially in the overcrowded areas, and the seasonal suitability of the various foodgrains in relation to the water supply. Moreover, the overcrowded areas contain not just a farming population pressing on the limited resources of each holding, but a large number of households who are either without land and must therefore seek employment as hired labour, or else who have only a little land and depend mainly on wage employment outside agriculture. In addition the towns draw on the rural areas for their food supply and for part of their labour supply, especially casual labour from the huge numbers of apparently unemployed. Thus one can no longer interpret the map of subsistence cropping in subsistence crop terms alone. The growth of towns, of industrial occupations and crafts, of service industries and the development of commercial agriculture are all taking place within areas which have hitherto been devoted mainly to subsistence crops. This commercial growth has in India not so much helped to solve the problems of pressure upon soil and water resources of a growing population as it has increased the precariousness of the situation by the frail support it has extended to a landless class, partly acting as rural labour and partly as a largely under-employed urban labour force, in some cases still supported by funds from the rural areas.

Subsistence crops dependent on rainfall have regional distribution patterns in which the constraints of moisture supply are less marked. Linear concentrations in river valleys and the crowding of deltaic and coastal lowlands are replaced by a normally more gradual moisture constraint gradient associated with the rainfall distribution, the constraints of soil distribution and the effects of culture and population distribution. Even in largely subsistence societies, where only small quantities of agricultural produce are traded, and service functions are severely limited, central places exist, providing centres of authority and frequently of defence. Often the crowding of population around such centres reflects not so much a wise choice of centre location with regard to soils, as the social advantages which are thought to derive from close contact with the centre, and may be seen in the crowding of small settlements around Kampala, the capital of Uganda, which today repeats on a larger scale the crowding that formerly existed around the traditional Baganda capital of Mengo. Even more spectacular is the crowding of people for some 30 miles radius around Kano in Northern Nigeria with a distinct decline in density as distance from Kano increases. Such factors produce distinctive patterns in the major subsistence crop distributions, including a tendency to 'clumping' in the distribution, that is grouping in local concentrations often with a very high density nucleus and some thinning towards the environmental margins. In Uganda, for example, the distribution of the staple plantain or cooking banana crop (Fig. 22 based on McMaster, 1962) is confined mainly to the Baganda and peoples of related culture. The Nilotes to the north are seed agriculturalists and the Nilo-Hamites both cultivators and herdsmen. Banana cultivation is strongly concentrated in five major districts: Ankole, Masaka, Mengo, Busoga and Bugisu. Four of these have a recognizable urban administrative core, including Mengo, which has a marked tendency towards higher density at the centre. McMaster concluded that rainfall was the main control

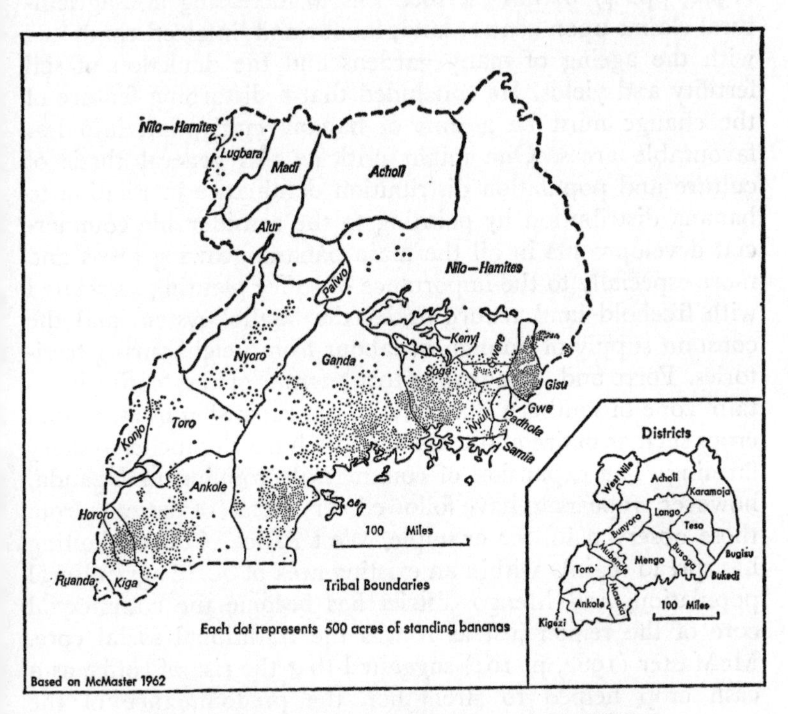

Fig. 22. Bananas, tribal areas and districts of Uganda

over successful banana cultivation, and in so far as cultivators have sought suitable banana land so rainfall must have exerted an effect on population distribution—'the people have clearly made effective empirical judgements as to land potential' (McMaster, 1962, pp. 22 and 44). Yet one can hardly ignore the concentration in the pattern. Ankole, Busoga and Bugisu are ethnically distinct from the two Baganda districts of Mengo and Masaka. The banana has been identified with Ganda culture, but also with the desire of some neighbouring communities to plant the crop as a means of identifying with that culture. There is some evidence that amongst the Gisu, for example, finger millet was the principal staple until recent times (McMaster, 1962, p. 28). There is also some evidence that population pressure is affecting the distribution. For example, in Mengo some decline in the banana area has been noted in the central counties accompanied by an increase in the outer counties. McMaster (1962, pp. 47–8) has ascribed this to increasing non-agricultural claims upon central land, labour and 'interest' combined with the ageing of many gardens and the depletion of soil fertility and yields. He concluded that a disturbing feature of the change must be a shift of banana cultivation into less favourable areas. One might criticize this general thesis of culture and population distribution dominance in relation to banana distribution by pointing to the considerable commercial developments in all the main banana growing areas and more especially to the importance of coffee planting associated with freehold land tenure under the 'mailo' system and the constant supply of immigrant labour from neighbouring territories. Fortt and Hougham (1973) wrote of the 'coffee-plantain' zone of southern Buganda, of the expansion of associated crops such as maize and tea, and of such developments as dairy farming. The expansion of commercial cropping in Uganda, however, appears to have followed very different patterns from those observed in, for example, West Africa. Coffee planting has spread mainly within an existing area of dense agricultural population, and Mengo district has become the commercial core of the region just as it was the traditional social core. McMaster (1962, p. 104) suggested that the rise of coffee as a cash crop helped to strengthen the predominance of the banana as a food crop because 'the two crops fit effectively together into a system of cultivation'. Thus a distribution pattern which is derived from relationships based largely on subsistence food crop production is still recognizable in an area very much affected by modern commercial developments.

Commercial crop regions: export crops

In most Third World countries commercial agriculture on any great scale is a recent development and a commercial agriculture whose practitioners depend either wholly or in large part on the market for their food supply and the supply of any other agricultural commodities which formerly they produced for themselves is a very recent development. Some Third World countries, for example Brazil, have long established traditions of export crop production, but usually by estate systems existing side by side with peasant subsistence farms, and confined to certain especially favoured regions. The development of commercial agriculture has taken place largely in response to the pressure of rising overseas demand in the late nineteenth and early twentieth century for tropical raw materials. The intervention of traders made small farmers aware of the considerable possibilities of the new market and bulking up the produce. governments provided many of the important elements in the essential infrastructure and also supplied information, advice and even financial support. With some crops, e.g. export bananas, traders or government agents have played a vital role because the risks of shipment had to be borne either by the growers or by their agents as far as the port of landing.

There are important differences in regional structures and distribution patterns between perennial crops and field crops (largely annuals), between export crops and crops for the internal market, between crops advancing into new land (colonization) and crops advancing into or through existing agricultural distributions, and finally between commercial crops which are 'primary' and dominate a regional distribution pattern and those which are 'secondary', i.e. which are planted in areas already dominated by a primary commercial crop and which benefit from the commerical developments that have already taken place. Most commercial perennial crop regions have an original core area, an area of current maximum production, which may differ from the original core area, and a pioneer fringe zone marking the frontier of production (Fig. 23). The original core area is usually as close to a major port, or to a frontier town of exit to a major port if the state is inland, as environmental constraints will permit. Sometimes, as in the older cocoa planting areas of West Africa, proximity to port is so important that the original core area is somewhat marginal environmentally and may quickly succumb to disease attack or

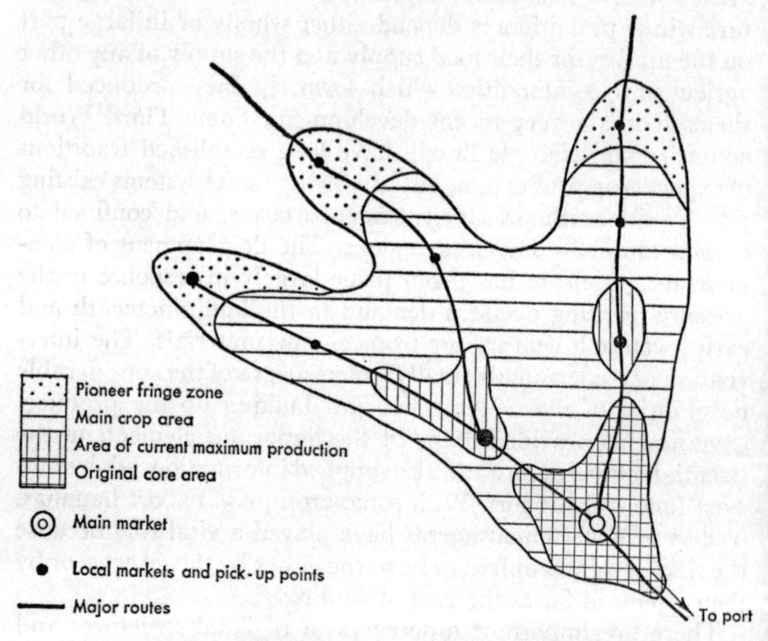

Fig. 23. Idealized perennial export crop region

to an over-rapid decline in yields. Such an area may be subject to rehabilitation, often involving the planting of new commercial crops, for few established growers will want to return to subsistence.

Often the pioneer research stations and seedling or seed supply centres are associated with the original core area and find themselves increasingly distant from the areas most needing their services, which eventually have to be provided through new out-stations. The area of current maximum production is the centre of crop dominance in the region, where all other crops are usually of less importance and where marketing of the crop and the return flow of consumer goods have their most intensive development. The less important crops should not, however, be ignored in any analysis. Often they provide most of the staple foods consumed by the otherwise commercially oriented farm population. Many Third World farmers have learned to depend on the market for their food but most, including many farmers otherwise highly commercial in their agricultural activities, persist in subsistence production, sometimes to minimize market risks, sometimes because of deeply embedded traditions (Upton, 1967, pp. 83–5). Characteristically the area of maximum production has important market towns, local service centres and pick-up points, and consists of a collection of nodal regions, each with a commercial focus, often linked by the major routeways as linear groupings of approximately circular (distorted by local, especially environmental constraints) areas of production. The most striking example is that provided by coffee production in Brazil, in which all three elements may be seen (Fig. 24 and pp. 215–20). The pick-up or buying point distribution may play a crucial role. If buying points are far apart production may be discouraged, as argued for cotton production in Western Nigeria (Kolawole, 1973).

As the advance takes place inland so the contact zone with either the wilderness or the subsistence production area tends to widen and the areal spread tends to assume the form of a rough triangle focused on a major commercial centre and with an outer curved edge, but a triangle broken into a number of routeway alignments each branching in turn into an approximately dendritic pattern. Major topographic barriers, the development of alternative seaports and routeways and the existence of major towns, provincial capitals which become secondary foci of major importance, all affect the distribution, producing considerable variations in the pattern. Increasing costs of transport with distance inland encourage the 'filling-in' of gaps in the dendritic distribution and the eventual occupance of most or even all suitable land so that the pattern may more and more assume the general shape set by ultimate environmental constraint. The area of current maximum production of the primary crop has usually, although not necessarily, the highest yields. It has the biggest concentration of farms with the primary crop and the farms with the highest proportions of land in the crop. It is often the main focus of innovation and agricultural improvement activity, more especially in the use of chemicals, new tools and techniques, since so many of its farmers are all interested in the same crop, although frequently the main area for planting the latest varieties of the primary

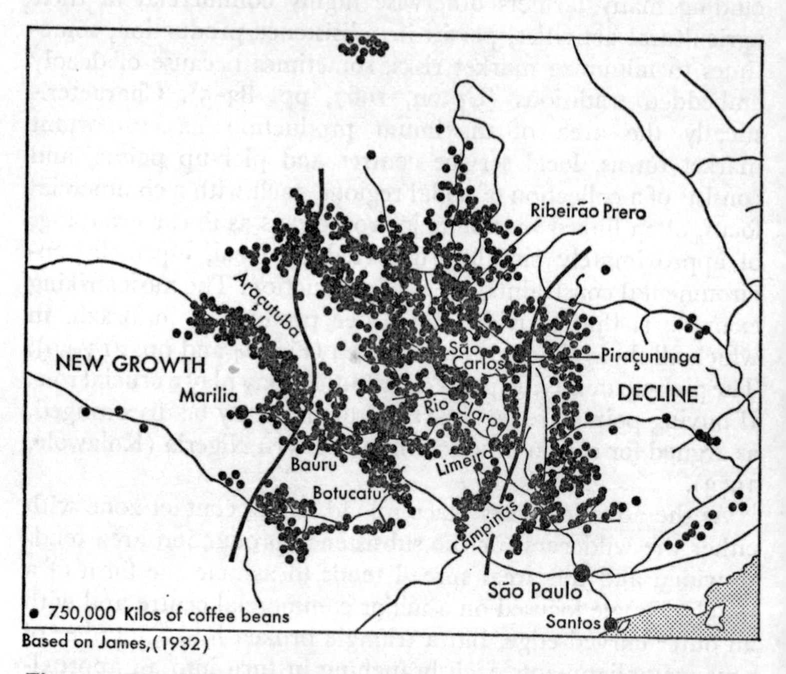

Based on James,(1932)

Fig. 24. Distribution of coffee production in São Paulo, Brazil in 1927–8

crop is in the pioneer fringe. Often, however, important innovative activity may take place in the 'gaps' within the core area which are also, in a sense, on the fringe of agricultural advance, although often better supplied with information and possessing better marketing links than the fringe. The pioneer fringe is one of scattered and sometimes low productivity as the new commercial crop is established by pioneer cultivators or leading innovators who open up new lands, sometimes just in advance of the arrival of the essential roads and marketing services.

Often pioneer cultivation can survive only by a mixed production system in which there is a large subsistence crop element or in which there are secondary commercial crops chosen to provide a quick return in a ready market although they may not be as profitable in the long run as the primary crop.

The contrast between commercial perennial and commercial field crops in their roles in farming systems in the Third World gives rise to significant differences in distribution patterns. Commercial field crops are normally combined with other field crops in a single system, sometimes even in mixed stands in the same fields and frequently in rotation with other crops, more especially with subsistence staples. Groundnuts, cotton and tobacco are thus found in general combination with grain crops such as sorghum, maize and pennisetum millet in rainland cultivation. When grown singly in fields which have no other crop they can occur as a distinct 'core cultivation' or as a shifting fringe cultivation to areas growing other crops. Most commercial field crops are the product of peasant farming systems with varying degrees of concentration on commercial production, many of which are in the process of becoming more commercialized. The result in geographical distribution is that commercial field crops both spread through existing farming systems, developing a new niche in them, and promote the expansion of the farmed area. Usually the farmers concerned combine commercial and subsistence crops and have farm organization and land tenure repeating traditional modes or modified forms of traditional modes. Where the proportion of each holding in commercial field crops is unusually large, as in groundnut production in Senegal, where a half to two-thirds of holding areas has been achieved, then areal expansion becomes an extremely marked feature and also includes a fringe expansion of the area in subsistence food crops, the amount of which on each holding may have been reduced by commercialization, even necessitating food imports. Senegalese groundnut production is in effect a special case, where close proximity to a port combined with especially favourable prices in the French world trading system and cheap food imports created a high inducement situation whereby groundnuts rapidly became dominant in the economy. The result was that although groundnuts combined well with food crops in existing farm systems nevertheless new groundnut dominant systems were created and planting was achieved more by areal expansion than by spread through an existing cropped area.

By contrast in Northern Nigeria, considerable distance from the coast and lack of especially favourable trading terms (although freight rates for groundnuts were low) reduced incentive. This created a situation in which groundnuts became an important cash crop, although on most farms they occupied a smaller area than staple food crops. Some expansion of planting has taken place but a large part of the groundnut crop was and still is produced on long established farms. Generally in the early stages of field commercial crop planting and for the most part in systems where less than about a third of each holding area is in the new commercial crop, most of the commercial cropping spread is within an existing agricultural region. In consequence the initial rate of spread may be very rapid and existing roads and marketing systems, however rudimentary, may be adapted to serve the growing commercial system. Fluctuations in profitability may result in crop substitution so that commercial field crop production is often very flexible and can tend to as rapid contraction when economic conditions are adverse as to expansion in the boom stages. Some smallholding farmers may even increase production as prices drop in order to maintain incomes, but many peasant growers will substitute other crops, especially food staples, when economic conditions are especially adverse. Where distance effects are an important element in costs (which depends on whether transport freight rates are adjusted to distance or, as in some cases, standardized), some tendency for distributions to 'pulsate' about a central core may be observed. Unfortunately data available seem too unreliable, coarse or fragmentary to test this. Few commercial field crops are grown in plantation or estate systems and those few are mainly the longer term crops such as bananas and especially those which need processing in or near the holding such as sugar cane. Even these are now also grown commercially on smallholdings.

In contrast perennial crop systems have mostly failed to fit existing systems of peasant farming. Coffee planting in southern Uganda, however, seems to have been an important exception, reflecting the successful combination of coffee with the existing food staple in a single farm system. One possible explanation for the special features of coffee distribution in Uganda is the considerable distance from the coast, which increased costs and reduced the profitability of coffee production by comparison with, for example, the Ivory Coast, so that coffee planting was not quite as attractive as in West Africa and did not rise to a comparable position of dominance in the agricultural economy. The contrast is comparable with that already described for groundnuts in Northern Nigeria and Senegal. Sometimes tree crops are an important but minor feature of smallholdings, producing mainly subsistence staples. Where these were already established, as with oil palm cultivation in West Africa, where adventitious palms are self sown in the fallows, a commercial tree crop production could spread through an existing, mainly subsistence smallholding region. For the most part, however, the needs of tree crop management, combined in the initial stages with high profitability and also the need to produce a certain minimum quantity to make dealing worth while, have combined to create farming systems in which all other cropping became subordinate. In these systems perennial crops required new and nearly permanent forms of land tenure, and mixed cropping was confined to the early stages of production only, and frequently as a cover or shade crop. Generally until recently, and sometimes even now, the most profitable course of action was to expand production by opening up new, preferably 'virgin', lands and abandon the old established areas as yields declined, when in effect they became worked out. Enormous areal spreads, usually into new lands, have been achieved and the main target of attack has been the tropical rainforest, most of which is now threatened by rising world standards of living and increasing demands for coffee, cocoa, natural rubber and palm produce. Whether plantation or smallholding systems have been involved, the effect in general or regional terms has been much the same. Tree crop systems have also proved somewhat inflexible, as they need several years to come into production, and growers, having sustained establishment costs, are reluctant to cut trees down before their many years of high productivity are finished. Booms are normally accompanied by considerable increases in planting, but slumps are normally accompanied by maintenance of the existing cropped area with a temporary halt in additional planting. Controls to 'even out' this process have been discussed above (pp. 125–7).

Commercial crop regions: crops for internal markets

Commercial crops for internal markets may be either perennial or field crops, and where grown on a massive scale may be expected to reproduce the kinds of geographical distribution discussed above. Mostly, however, they are not grown on as massive a scale as export crops nor has their production expanded so rapidly. Instead they have tended, subject to environmental and market constraints, to be more widespread in their distribution, that is somewhat less regionally specialized, and to expand slowly in somewhat localized spreads, sometimes around major cities, sometimes on the fringes of extensive export crop areas creating local demands for agricultural produce, sometimes even within such export crop areas and sometimes round processing centres, such as a fruit juice canning factory, oil palm fruit crushing mill or cotton ginnery. Commercial crops for internal markets are extremely varied and exhibit correspondingly varied patterns of distribution.

In practice most situations are mixed, that is there is a considerable overlap of the various commercial and subsistence systems, usually with the result that some crop is dominant and the rest secondary, although occasionally with the result of co-dominance or a situation in which it is difficult or impossible to establish which of two, three or even four crops is the most important in the farming system. Often it is difficult to be sure which role is intended for a particular crop, especially with food grains which may be intended for domestic consumption as flour for porridge or bread, or for malting and beer, or for sale, usually as grain, to an internal regional or national marketing system, or for export abroad. Some grain crops may achieve more than one role, part being sold and part being retained on the farm. Some commercial crops in their expansion create in effect a new distribution pattern for the secondary crops which follow in their wake or accompany them even in the earliest stages of expansion. Thus cover and shade crops may become an 'extra', even a 'surplus' source of timber and food and the main subsistence crop or commercial food crop area may move outwards together with the main area of concentration of pioneer cultivators, especially in a perennial crop region where less labour is required in the later stages and therefore earlier locations of production. Where labour remains behind, it has often turned to market gardening around the growing towns of a new commercial infrastructure.

The pioneer fringe areas of commercial crop production are often very special areas of food cropping and may even have contained precursor or pioneer food crops planted before the invasion of commercial tree or industrial crops, perhaps for subsistence in the early stages of settlement establishment or for sale as commercial food crops in the towns of the expanding tree or industrial crop region. Sometimes, however, commercial perennial crop expansion is not associated with food crops either as shade or as interplanted crops, as, for example, in the expansion of some forms of plantation production, or where preferred shade crops have little use other than as firewood, as in the use of *Gliricidia* spp. ('mother of cocoa') as a shade crop for cocoa in Ghana. Where commercial crop expansion in such areas is very rapid the import of foodstuffs may be essential and well served by the return of otherwise empty vehicles on the transport arteries built for export traffic.

The kinds of regional combination and overlap that may obtain can be illustrated by an attempt to model the situation in a tropical area with well developed tree crop production (Fig. 25). This diagram is an empirically based attempt to describe the kinds of spatial pattern that occur, more especially in regions of mixed peasant agriculture dominated by the production of staple food crops and by commercial smallholdings devoted mainly to tree crops for export. The pattern suggests a broad contrast between a 'developed' or commercial core area and a more subsistence, inward looking periphery, but does contain a number of complicating elements. Firstly, there are not two zones but four:

a. a zone of tree crop production mainly for export which also includes food crops, often somewhat extensively grown because labour resources are diverted either to the nearby towns or to commercial crops, with the exception of some intensive market gardening, often affected in less developed countries by a distance function;

b. an inner zone adjacent to urban areas of investment in farms as farm property and also in farmland as future sites for urban development, and including some intensively farmed land either as market gardens or in effect as allotments or very small holdings worked by people who also have jobs in the towns;

c. a peripheral zone of export crop expansion, usually in road oriented salients and 'islands', and of commercial food crop production and industrial crop production;

d. an outermost zone of subsistence farming with some

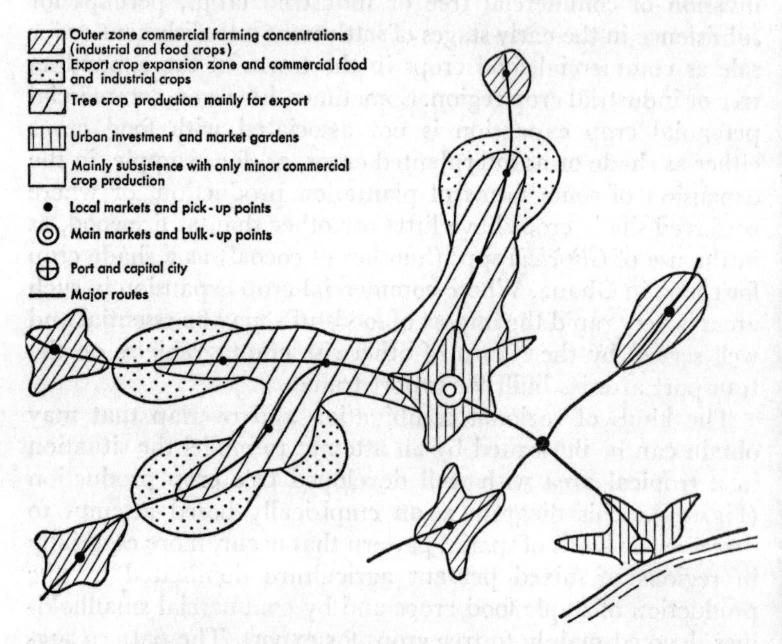

Fig. 25. Idealized combination of commercial tree crop, commercial food and industrial crop and subsistence production regions

commercial farming in accessible locations, but often possessing an extensive character, i.e. using low input shifting agricultural methods on small farms or mechanized methods on very large farms.

Sometimes limited road networks result in poor commercial accessibility in rural areas quite close to large cities, and an apparently 'wild' landscape in which scattered farms exist, using supposed primitive shifting agricultural techniques, occurs close to the city. The key element in this pattern is undoubtedly the region of tree crop production mainly for export in that it is the pioneer developments in this region that have initiated the chain of changes which have spread to all regions and it is frequently this region that is the major source of taxable wealth used to foster varying developments in the other regions. In effect this exemplifies export-based growth as discussed in a rather different context by Baldwin (1956). The idea that such a region is a distinct 'enclave' of modernization having few relationships with other regions, except perhaps in so far as it drains their labour resources, whilst it contains many elements which fit the observed situation, is a dangerous view of processes which, however tenuous, nevertheless are creating a complex regional system far more extensive in area than the so-called 'enclave'.

Something of this regional complexity may be glimpsed in two papers describing West African conditions and in a recent study in Uruguay. Engmann's (1973) study of spatial convergence-divergence using 25 selected population variables in Ghana resulted in eight development regions, whose locations and agricultural characteristics, modified by recent decline factors arising from disease and ageing plant stock, suggested the kind of pattern described. Ajaegbu's (1970) study of food crop farming in southwestern Nigeria indicated a distinct zonation of agriculture from an urban related zone of vegetables and fruit, through a 'plantation' crops zone to a zone of cassava and maize including commercial 'sole' crops, finally to an outer zone of lumbering, firewood and subsistence. A similar indication was provided by Griffin (1973) in the very different environment of Uruguay with his Thünen arcs of horticulture, dairying, cereals and cattle grazing. Griffin noted that the distribution of the individual land use regions did not conform to the ordering of intensity classes anticipated in the theory, but Thünen was careful not to order his zones as intensity classes and to allow in some respects for innovation—hence the not so unexpected fair relationship between the early Thünen model and the patterns created by the spread of different forms of commercial agriculture in the Third World countries.

In Zambia, by contrast, commercial cropping for internal markets is not associated with export crops but with mining. Baldwin (1966) has questioned the apparent general failure of the rapidly growing export sector to generate much spread effect through the economy as a whole and has noted that most benefit has gone to only a small part of the population with very little development in the agricultural sector despite its readiness to meet the demands of industry for labour and farm produce. Yet agricultural zonation around the major mining and industrial areas is apparent, with wide rings created mainly by the low density of population. Baldwin noted an inner zone of vegetable production, a firewood zone at about 80 to 100 km and a 'weight-reduced' crops zone beyond 100 km. Some produce came from as far away as 1200 km from market (Miracle, 1962). The general suggestion in these examples is of areas of declining intensity of production outwards, but this is not necessarily so where different products are concerned or where other variables complicate the situation. Outwards from the city of Ibadan in Western Nigeria, for example, the author has observed some intensification of yam production: farmers near the town either concentrate on other crops or lose family labour to the town. In consequence much hired labour is used for food crop farming, often unused to the special techniques of intensive yam cultivation. In any case hired labour is expensive and labour costs are reduced as much as possible. Lower yields are acceptable where inputs are reduced, usually by making only low mounds and not erecting frameworks of poles and string for the vines to climb. Further out, yams are produced on family farms, where every effort is expended to achieve high yields, more especially because yams are normally planted on the small areas of most favoured, moister soils (Morgan and Moss, 1970). A form of 'middling' input technique, involving some extra effort and higher yields, found at some distance from Ibadan, is the use of guinea corn stalks to produce a yam climbing frame, guinea corn having been planted before yams in a rotation. This is not necessarily because of labour problems, but because of local high costs in obtaining yam poles (Moss and Morgan, 1970).

Livestock regions

The treatment of the regional organization of livestock rearing in the Third World has been separated from that of crops partly because there are many areas, more especially in Latin America and Africa, where livestock occupy distinct regions or where associated cropping occupies a distinctly secondary role, and partly because even where livestock are closely linked with cultivation, in the tropical world mainly for draught purposes, special considerations make separate treatment useful. Although some regions of livestock production are entirely commercial in character, the existence of purely subsistence livestock regions is very rare. Most of the livestock regions regarded as traditional have a measure of both commercial and subsistence organization. In practice little was to be gained by an attempt at separation. There is a fundamental ecological distinction between the tropical grazing lands such as the drier African savannas and the more temperate grazing lands such as the Argentine and Uruguayan Pampas. Pastures in the latter, whether sown or self sown, deteriorate less quickly, are generally more nutritious, suffer fewer disease problems and are more suitable for breeds of livestock imported from Europe, for whose meat there is a ready market overseas. In consequence in the Third World livestock regions the development of overseas exports has been confined mainly to the few more temperate lands, more especially to Argentina and Uruguay, and has been slow to develop in the tropics, where commercialization has looked mainly to inland markets. Whereas in the temperate lands one of the most important regional features has been the growth of huge areas of dairy cattle concentration, such areas are largely absent from the Third World, partly because of differences in development and of the limited markets for dairy produce, and partly because so much of the Third World is tropical, with all the problems of treating and preserving dairy produce that that implies.

India has the largest concentration of cattle in the world, with probably about 200 million head, whose geographical distribution is almost identical with that of the population. In effect India does not possess a distinctive cattle raising region, only a general aggregation of cattle, used, if at all, mainly for draught purposes and to some extent for milk supply, together with water buffalo used to plough the rice fields. Sheep and goats by contrast are distinctly regionalized in a large concentration in the drier lands of the northwest, and combined with cattle in southern India, chiefly in Madras. India's overwhelming cattle population is the result of Hindu beliefs which forbid the eating of beef and also a reluctance to slaughter the older animals. The land provides very poor grazing with limited areas of village common lands and straw and other crop refuse. Huge numbers of cattle are useless and most suffer from malnutrition. Milk production is low even for the limited demand and many farmers lack adequate draught power despite the cattle surplus. Over half the milk consumed comes from the much smaller buffalo population. The chief concentrations of dairy cattle are around major cities such as Calcutta and Bombay, and in the richer agricultural areas of the irrigated lands of Punjab and Rajasthan. The location of sheep raising is governed largely by the avoidance of lands suitable for agriculture, that is, sheep are confined to marginal, more especially semi-arid, pastures and hill lands.

In tropical Africa cattle, sheep and goats have broadly coincident distributions in the semi-arid fringe lands of the cultivated savannas. Livestock rearing here depends mainly on nomadic pastoralism, either in elongated north-south movements in southern Africa, West Africa and the Rift Valley lands of East Africa, or between uplands and lowlands. Movement is essential, for suitable grazing on the open range is short-lived. Young grasses provide the best fodder and burning is frequently employed to remove old coarse material. The lopping of trees provides supplementary fodder, and floodland pastures play a vital role in crucial dry periods during the seasonal movement. Regionalization is linked to ethnic groups such as the Fulani of West Africa and the Nilo-Hamitic groups of East Africa and to key water points, salt licks, taxation points and 'bases' often with some cultivation. Distinct territories may be generally recognized but their boundaries are frequently adjusted. For the Somali of the eastern Horn of Africa, for example, ownership of the unenclosed land is tribal or the common property of the people as a whole, to which no individual or group may claim an exclusive right (Allan, 1965, p. 295). Although the development of commercial livestock raising is taking place elsewhere in tropical Africa, these semi-arid lands are still the main source of meat and hides, and often the former pastoral nomads have become commercial meat suppliers to the cities of the agricultural lands. Some, chiefly the pastoral Fulani, who move their herds between upland and lowland pastures, have been encouraged to become dairy farmers, selling milk to dairies established by the government, but most walk or more often nowadays transport their cattle by road or by railway to the commercial centres of the cocoa and coffee producing regions or to the capital cities. In the Teso District of Uganda milk collecting points have been established at local administrative centres with an effective collection radius of about 15 km, the distance beyond which most movements appear to need transport for which cash payments must be made (Funnell, 1973, p. 8). Attempts have been made to create more permanent pastoral settlement by pasture improvement, controlled grazing and the provision of better water supplies, mainly from bore-holes. The pastoral regions were extended by creating ranching schemes and fattening pastures in the moister savannas in locations nearer to the major meat markets.

Many governments regard nomadism as anachronistic in a modern state and are endeavouring to eliminate it. In some instances, notably in Middle Eastern countries (George, 1973), they treat it as a national problem (which should logically have been discussed in the previous chapter). In Egypt, for example, the Desert Law, enacted in 1958, considered tribal lands as state property unless proof of planting and development was provided. Schemes such as the New Valley Project, which involves irrigation of oasis lands in the Western Desert, will be an inducement for bedouins to settle. In Syria settlement of the nomads became a basic tenet of policy in 1953. Cereal cultivation has been encouraged on land formerly used as pasture and attempts are being made to promote a system of range management linked to permanent villages. Mixed farming techniques have also been introduced, mainly aimed not so much at an equilibrium between cropping and manure supply, which is difficult to achieve, but at the introduction of oxdrawn ploughs in order to eliminate the tillage bottle-neck and make possible larger farms (Uzureau, 1974). Some success with such mixed farming has been achieved wherever tillage is an important bottle-neck in production, more especially in Zambia and Rhodesia, in Uganda, where many cultivators already own cattle, and in Mali, where ploughing is applied to floodland. In Northern Nigeria success has been much more limited despite nearly a half century of effort. Frequently the key problem, apart from disease, which is an especially serious threat in Africa, is the provision of fodder all the year round without migration. Where output per man is low peasant farmers are not anxious to grow fodder crops (Floyd, 1959, pp. 198–9 and Morgan and Pugh, 1969, pp. 508–9 and 513).

In Latin America comparable use of savanna lands for livestock rearing occurs mainly in the Orinoco *llanos* of Venezuela and in the *campo cerrado* of the pastoral *sertão* of northeast Brazil beyond the *travessão* or legal line of separation between open range and fenced land, and in Goias and the Triangulo of Minas Gerais, where grazing has been stimulated by the extension of road and rail communication to the industrial cities of the southeast. Special fattening pastures have been provided to improve the quality of meat sold, notably at Barretos and Uberaba. In southeast and southern Brazil livestock rearing, particularly of dairy cattle and pigs, has expanded considerably in recent years in response to urban influence and, in some areas, as a form of diversification following the decline of former boom crops such as coffee, oranges, cocoa and sugar cane. The maize-pig-cattle combination appears to have had its origins in the more temperate climate of the semideciduous forest lands of the south, where European colonists, mainly of German origin, created a successful mixed farm economy, including also rye and potatoes, not dissimilar from that created by their counterparts in the United States Corn Belt. A regional structure which in its inception was in part based on subsistence has now become entirely urban market oriented. The use, however, of fodder crops in livestock rearing and of other modern inputs is still very limited in Brazil, where most livestock depend on crop refuse or poor quality self-sown pasture and where hand labour using hoes is still a significant factor in production.

Pastoralism dominates the life of the Humid Pampa of Argentina, making it one of the most productive livestock regions in the world. Despite the importance of access to export markets, the distance factor, that is access to the market of Buenos Aires, has had only a small effect on zonation within the Pampa, where the economy of managing its varied environments has largely dictated the distribution pattern. Thus the extremely humid environment of the eastern portion with its cool, moist summers is less suited to grain crops than to grass production and over 80 per cent of the land is in pasture, chiefly for beef cattle and for high quality mutton and wool sheep. Immediately around Buenos Aires specialized dairy farming and market gardening in a zone of some thirty miles radius, reflects the accessibility effect, but the alfalfa-maize-flax zone to the west and north of Buenos Aires and the alfalfa-wheat zone (both including livestock), which occupies most of the Humid Pampa in a great arc from Santa Fe in the extreme north to Bahia Blanca in the south, reflect contrasts in environment whereby each grain outyields the other in its appropriate zone; maize gradually replaced wheat in its zone after 1895 as the maize export trade grew through Rosario. Most of the maize is grown by tenants, chiefly of Italian origin, on small farms, whilst the remaining area of the estates is devoted mainly to alfalfa pastures and livestock. Similarly tenant farmers account for most of the wheat production on estates otherwise devoted to livestock, although there is a growing trend towards mixed grain and livestock farming. In Argentina the pastoral region has a much more varied character than elsewhere, is everywhere export oriented, including all its major agricultural enterprises, and has been dominated by the commercial foci of Rosario and Buenos Aires and the transport web. Pastoralism is in no sense a fringe activity or restricted to marginal lands, and it has proved capable of earning high incomes, although income distribution shows a great range of earning capacity. A capitalist class has emerged whose investments have proved an important link between town and country and a means of moving capital between the agricultural and industrial sectors. Livestock rearing in the Pampa has proved not just a useful pioneer stage towards development, but a continuing source of economic strength. Changes in the European economy and more especially the development of the Common Market threaten the future of the livestock industry.

They should encourage more crop based farming systems, possibly of a more intensive kind, and more investment in industry.

AGRICULTURAL REGIONS AND AGRICULTURAL INPUTS

The attempt to describe agricultural distribution patterns and to type them has shown not only the expected important role of market, that is demand location, in the evolution of commercial regions, but has also indicated the importance of input factors. The availability of land as a determinant of distribution pattern has been seen to be important in the spread of coffee and other commercial crops in Brazil, where the *sertão* or frontier zone has played a crucial role in the evolution of the economy. Similarly with cocoa in Ghana and the Ivory Coast and with groundnuts in Senegal the existence of huge empty areas into which the new cropping could spread helped to create situations of commercial crop region evolution in which other crops have become of secondary importance in the pattern. It also encouraged a low input situation, because of the cheapness of land, in which low costs have more than compensated for low yields or for yields which were high only because of the inherent fertility of freshly cleared areas. Inland advance has inevitably resulted in increased costs of transport not only for the evacuation of produce but also for imported goods. These have reduced contacts with commercial centres and heightened the difficulties of obtaining information and supporting services. Inevitably as such costs have risen so the alternative of more intensive production has looked more attractive. In Senegal the potential area for groundnut planting was extended by bore-holes, intended originally to supply water for cattle raising (Morgan and Pugh, 1969, pp. 509–10 and 639), but expansion has been overtaken by slow growth in the world demand and a virtual collapse of the favourable prices of the French commercial system. In Brazil the occupation of empty space has become a major political policy, and new growth poles have been built in the interior, creating new distance-demand effects. New patterns, however, result because the commercial export crops still face the high costs of increasing distance to port, and the interior regions depend much more on ranching and food cropping for the internal market. The huge growth of cities like São Paulo, Rio de Janeiro and Belo Horizonte has increased the demand for commercial foodstuffs, particularly rice. The interior state of Goias has benefited from this with increased 'exports' of rice and also of beef cattle to the urban areas. The intrusion of Brasilia with its 400 000 population has not, however, stimulated further outward expansion of agricultural growth, but has merely diverted an existing growth in response to a national urban demand trend to a new location (Katzman, 1975).

There has been for some farmers a transport cost reduction, almost certainly more than matched in regional accounting by the increased administrative costs, accessibility costs and capital costs of the new city. It is worth noting that the building of interior cities appears to have done little to stimulate local agricultural production. In Goiania, the state capital of Goias, established in the late 1930s, Katzman has claimed that man-land ratios were not high near the city, but were high on more fertile land in the Mato Grosso, that is labour substitution for land was governed much more by environmental differences than by location with respect to market. The contrast of a land shortage situation may be exemplified by Egypt, where high capital expenditure on irrigation works has been tried to effect both an increase in the cultivated area and increased yields, i.e. a capital substitute for land. Through these measures the cultivated areas rose by nearly 20 per cent between 1897 and 1960, but the cropped area, with double cropping, rose by over 50 per cent. In the same period, however, the population increased from an estimated 9·7 millions to 26·6 millions and the cropped area per head fell from 0·3 hectares per capita to 0·16 hectares. The cropped area expansion may be seen therefore as part of the general demand for land, so that the growth of commercial cropping, more especially the expansion of cotton production during the period, although it undoubtedly helped to pay the costs of irrigation and increased the money yield per hectare, may be seen as a spread through an existing agricultural distribution pattern whose total expansion was in effect a per capita contraction. Cotton was especially favoured as the leading commercial crop of Egypt, partly because as a summer crop it avoids the winter period favoured for wheat and barley (although it conflicts with rice and sugar cane which often replace cotton in Upper Egypt), partly because the crop is inedible and will not be consumed by hungry tenant farmers in times of food shortage, partly because there is plenty of labour available and partly because Egyptian conditions are especially suitable for some of the world's highest quality long staple cottons, controlled mainly by government issue of seed. Cotton cultivation has in part replaced food crops, but has permitted the import of foodstuffs, especially wheat, in far greater quantities than could have been produced on the same area. In this connection Egypt's production and export of rice in order to pay for food grain imports should be noted. The Egyptian agricultural area is dominated by commercial crop production, but the expansion of this production has been by intensification, especially by the application of water and fertilizers, by double cropping techniques and by a reduction in the proportion of land devoted to subsistence crops, more than by areal expansion. The result has been the substitution within more or less the same area of new agricultural systems for the earlier food staple based systems. The new systems support far more people but depend on expensive techniques of water control and on the fluctuating prices, beyond Egypt's control, of a few agricultural products on the world market. The agricultural region has evolved mainly through a population pressure which has made land a scarce resource, thereby encouraging the substitution of high value commercial crops for lower value crops which have had to be imported.

Labour shortage has restricted agricultural development in some regions or has resulted in the decline in some forms of production as others have expanded. Yet it can also be argued that the high inducement of some forms of commercial cropping has served to mobilize labour. In Uganda, for example, McMaster (1968) has claimed that cash crops have done most to bring men more fully into agriculture. The expansion of commercial tree-crop cultivation in tropical lands involved a movement not only of farmers and their families into new lands but of migrant labour, both permanent and seasonal. Areas of limited agricultural potential, particularly with short rainy seasons, say four and a half months or less, and little or no floodland, particularly in Africa, have tended to lose labour to better endowed areas and chiefly to the long rainy season perennial crop producing regions of the former rainforest lands. In Latin America special groups of pioneer tenants have emerged, planting tree-crop seedlings intermixed with food crops and moving on once the plantation has been established. In southern Brazil coffee is grown on family smallholdings or *sitios* and here seasonal workers are employed mainly for the short harvest period. In northeastern Brazil short term tenancies occur on the large cattle-raising estates of the *agreste* or drier agricultural fringe lands. Smallholdings are leased for a few years mainly to grow food crops, thus extending the cropping region into the *agreste* zone. Cattle occupy the fallows subsequent to cultivation and a sort of symbiosis of cattle and cropping, tenant and rancher, is achieved with supplementary incomes for the tenant farmers from seasonal work on the sugar plantations of the coastal region (Sternberg, 1967).

In Africa temporary cultivators have appeared in some cocoa and coffee farming areas, but other groups have also appeared, including more permanent share-croppers and seasonally hired labourers, who have often assisted by tending not the perennial crops but the food crops, both for the farmers' families and for local sale. Thus the emergence of a regional system of export crop and food crop production has been accompanied by the movement to different parts of that system of various classes of migrant labour. As the system expands less fast and intensification has become more significant, so the migrant labour force has become more settled and the pace of agricultural migration has slowed. Urbanward migration has begun to take its place, sometimes with the regional agricultural system acting as a sort of routeway for migrants who have moved to villages and temporary agricultural work near the large cities before moving to the cities themselves. The general tendency to use gravity models to describe such migration is in accord with the tendency for greater movement to take place the closer the location lies to the centre of attraction. Nearby locations are better informed, usually have more contacts in the town than more distant areas and may develop better expectations of urban employment through greater and more reliable information. Todaro (1971) has developed a useful formula for calculating the discounted present value of the expected 'net' urban-rural income stream over the migrants' time horizon. Hart (1973) has argued that urbanward migration in part reflects a wage-earner's desire to diversify his income by trying to obtain several jobs. The general coincidence, apart from Zaire and interior Brazil, of rainforest and coastlands and the resultant proximity of tree-cropping to ports, has helped to reinforce this urbanward and coastward labour trend which has moved against the old frontier type advances. The modern development of interior growth poles may be seen not as a logical extension of earlier advance, but as a counter to current trends.

Attention has already been drawn to the considerable international labour movements which made possible plantation regions such as the rubber growing region of West Malaysia (p. 163). Within regions of varied crop production labour supplies differ between locations and have begun to affect crop location. Thus the pioneer fringe zones of commercial crop expansion may look like regions of opportunity to neighbouring areas of subsistence production and be able to attract labour, whilst the core areas of commercial crop production tend to lose labour to the cities. Such core areas have been affected not only by declining production from ageing perennial crops but by the loss of young people to the towns and an inability to attract hired labour except on a very short-term basis as a step on the route of urbanward migration. With often an ageing farm population it is hardly surprising if labour-saving crops such as cassava are sometimes preferred in such locations, unless highly profitable vegetables to be sold in urban markets can be grown, producing an income equivalent to the incomes that can be earned in the town. The emergence of cassava dominant zones close to some West African cities, with crops such as yams requiring more intensive effort grown at more distant locations, or grown by more extensive techniques, nearer the town, as discussed above (p. 192), may well be the product of the labour situation rather than of transport costs and some economic rent function.

Capital as a factor in regional evolution is much more difficult to measure and judge. For some commercial cropping developments very little capital has been required. Tree-cropping, however, even on smallholdings has required capital to survive the period before the trees come into bearing, to pay for seedlings and in some cases for chemical sprays and fertilizers, to pay for hired labour and to pay for equipment used in processing or preparing the crop before sale. Often perennial crop expansion has had to be achieved by the purchase of land, as long-term cropping of this kind involves virtually 'permanent' occupance. Much of the capital for such expansion today comes from the farmers themselves, looking for further expansion of their business, but some comes from banks or from agricultural credit institutions. In the early stages of development, mostly in the late nineteenth century, there were few wealthy smallholders. Estate owners in Latin America were able to expand their systems fairly rapidly, and the expansion of tree-crop cultivation in Latin America was generally ahead of that in Africa and Asia. Much of the profit derived from it has gone, however, not into further expansion or improvement but into industry and services or into conspicuous consumption. In Africa and Asia some of the capital for commercial export crops came either from overseas or from traders in the major ports. Chinese businessmen in Malaysia and Indonesia invested in small-or medium-sized holdings. In Western Nigeria traders helped to promote agricultural colonization (Akintoye, 1969) or bought land and invested in cocoa planting in association with other members of the professional class and in societies such as the African Church (Webster, 1963). In Northern Nigeria Hausa traders gave financial assistance to groundnut farmers (Hopkins, 1973, p. 220). There were commercial nuclei in the West African rainforest zone for the major tree-crop regions, orienting them towards the major ports, even though some of these early locations, such as the early Ikeja cocoa plantations in southwestern Nigeria, were environmentally marginal. Often the beginnings of commercial cultivation may be seen in low cost developments whose opportunities were appreciated by a particular social or ethnic group. In Western Nigeria the 'ethnic' community acted as a source of goods and services for cocoa innovators so that a traditional institution facilitated the exploitation of new economic opportunities (Berry, 1974).

Commercial cultivation in the Malay peninsula dates back

at least to the late eighteenth century, when Chinese immigrants began to cultivate in Kelantan and grew pepper by shifting cultivation techniques in Trengganu and later in Singapore and Johore. By the late nineteenth century some 200 000 Chinese immigrants were involved, but the cultivation declined after 1910 as rubber prices rose and as the shifting techniques used and the spread of smallholdings generally were constrained or forbidden by legislation. Roadside smallholdings, for example, had been forbidden in 1905. Successful pepper farmers invested in the much more capital demanding rubber plantations instead (Hill, 1969), where previously only the wealthier and more enterprising Chinese had invested (Voon Phin Keong, 1967), or looked to investments in mining and trade, where most Chinese capital was deployed. In the early 1950s high pepper prices encouraged fresh planting by the Chinese, but with an emphasis on labour saving spurred by increased labour costs. Plantations were established not in fresh forest clearances as in the past, but in more easily cleared areas of secondary growth where lower nutrient levels in the soils were compensated by the use of fertilizers and groundnut waste.

REGIONS AND INFRASTRUCTURE

The regional organization of agriculture in relation to infrastructure, which in less developed countries consists mainly of markets, transport services and systems, extension and information services, and the supply to farmers of agricultural tools, seeds and other inputs together with consumer goods, is often treated quite distinctly from crop and livestock regions. The latter are regarded as formal, the former as functional, which may be intertwined by reciprocal and complementary processes (Manshard, 1974, pp. 18–20). Here no such distinction is made. Subsistence and commercial crop regions are regarded as two extremes of a regional spectrum in which the degree of commercialization involves so-called functional regional structures, which mostly do not coincide with one another and which in combination make up a major commercial crop region. The emergence of a regional structure in which urban fields are major elements through the supply of services is not a new phenomenon in the Third World. Internal trade in many Third World countries was developed long before European imperialism made its impact, and the activities that took place in many rural areas were governed by decisions made at some central location. What is different is the degree to which commercialism has begun to affect rural areas, combined with the growing importance of the export sector and attempts at the national level to plan and direct agriculture. The regional structures that have emerged affect not only a particular commercial crop, but many other agricultural enterprises in the area concerned. This occurs directly through the operations of the agricultural planners and indirectly through the interrelationships of the enterprises concerned and the degree to which successful commercialization of one crop may encourage commercial experimentation with others. In some areas, as Preston (1973) has shown (and see p. 152), the town may be of little or no importance in effecting agricultural change. In others it has played a key role, being both the product of agricultural production and exchange and a cause of agricultural development through the influence of its markets and services. In the latter case it may be identified as an instrument of rural modernization, even Europeanization, and regional issues may be confounded politically with social and imperial issues. Thus in the Khmer Republic in April 1975 the evacuation of Pnom Penh and other towns ordered by the victorious communist Khmer Rouge was an attempt to destroy a commercial and supposedly colonial regional system in order to create a more isolated and subsistence society less subject to foreign influences. The importance of service provision may be judged in Ghana, where in addition to extremely high taxation of cocoa sales, farmers have been discouraged in recent years by poor provision of chemical sprays, the inadequacy of road building and repair programmes, and a shortage of vehicles to transport the crop. Production fell from 557 000 long tons in 1964–5 to 348 000 long tons in 1973–4 with declining yields and a static area, whilst elsewhere, notably in Brazil, production increased.

In most Third World countries the internal market system, both local and inter-regional, has generally involved far more people than the export sector. Although it has not been the means of effecting changes in production as rapidly as have the export markets, nevertheless it has provided local stimuli to production and meeting places for the exchange of information, even possibly for the introduction in the past of new crops. Since its spatial structure has not been determined solely by the needs of agriculture, but rather by social and political factors, the traditional market system may be regarded at least in some small part as having shaped the spatial structure of agriculture before modern commercialism developed. Most trading in the Third World takes place person-to-person in small, usually periodic markets, each serving a limited area, with a population of 3000–15 000 (Bromley, 1971). Daily markets are usually located in major market centres, often in towns. Special markets or fairs occur annually or at some other long interval. Hodder has argued that markets engaged in trade from outside the region and usually visited by caravan traffic were the first kinds of market to be established, that markets can never arise from the demands of purely individual or local exchange as such, and that such exchange occurs as a later stage in the development of a market system and hierarchy (Hodder, 1965).

The importance of external trading in agricultural development has already been stressed and its growth has undoubtedly stimulated other forms of production and exchange, including production for internal markets. There is, however, some tendency to dichotomy in the market system. Export crops are usually collected at pick-up points which serve as places for the bulking-up of goods, whilst produce for local sale is often handled through traditional periodic and daily markets. Some internally marketed produce goes to market or to wholesale and retail outlets directly through dealers, and some, such as tobacco, fruit and vegetables for canning or freezing, and cotton, are sold directly to processors or manufacturers, often under contract. Some periodic markets are not located in villages but in nodes, each accessible from a group of villages, often in the open countryside as in the Moroccan *suqs* (Fogg, 1932) and also the Yoruba markets found in southwestern Nigeria. A great deal of time is spent in trading activities in the Third World, but this is hardly surprising in view of the uneven demands on labour time made by agriculture and the existence of 'spare' time even where labour is short. Where incomes are low and sales involve bargaining the small losses and gains that occur are of vital concern. Most farmers and their families engage in trade, so much so that with many it is virtually a second occupation. The proliferation of small-scale intermediaries in Third World marketing chains is to be expected, together with break of bulk down to extremely small quantities to suit very low incomes. Forman and Riegelhaupt (1970) in a study based on Brazilian evidence have proposed an evolutionary model involving the lengthening of market chains and the proliferation of intermediaries until a point of growth is reached in which wholesalers can achieve scale economies by using higher capital inputs and higher turnover in order to undercut and by-pass intermediaries. In Bolivia we have a contrast in marketing development, where formerly bulk sales of goods were dominant under the estates system, but after the land reforms of 1952 small peasant producers sold part of their crops to wholesalers in the local markets, where they also purchased consumer goods (Clark, 1968 and Preston, 1969 and 1970). Many periodic markets are recent in origin and have been brought into existence as marketing systems have expanded or as new urban, industrial or mining demands for agricultural produce have been created. In Sierra Leone, for example, the periodic market is a recent feature, often large and attended by traders who travel long distances (Riddell, 1974). In the Punjab the size and spacing of the legally regulated wholesale markets dealing mainly in wheat were not related to those of the central places in which they were located. Communication facilities were the main locational determinant, whilst agricultural wholesaling links were determined not so much by competition between towns as by competition within towns between middlemen (Harriss, 1974).

Market development requires a transport system. Traditional markets were served by simple local systems, which cost little to provide or to maintain, but which could serve only a limited traffic. In many areas these have proved a constraint on economic advance. Such systems are for the most part ill suited to serve the needs of modern urban markets or export trade, although some remnants remain and many of the ancient routes have been incorporated into modern roads or railways. Modern transport, however, requires capital, and the major routes need a measure of central planning and direction, even though modern feeder roads may be the result of local effort and initiative. In consequence government transport policy may control the development of agricultural regions and marketing systems, either as a constraint by a failure to provide the transport needed for expansion to take place, or as a stimulator to growth by opening up new regions.

Marchand (1973) has argued that the transport systems of developing nations represent unique kinds of transformation. He cites the Venezuelan development in patches following a government investment policy which resulted in the co-existence of a small town pattern with a low level central place network, evolved largely in the nineteenth century, and of a middle-sized and large city system of more recent origin. In this situation urban growth appeared to have very little relationship to population potential. 'Demographic growth in such countries is not related to the commercial influence of the cities, but to the crisis of the agricultural sector, the breakdown of the traditional economy, and the powerful attraction of administrative activity.' Undoubtedly there is a need in most Third World countries for greatly improved transport networks and a greater spread of urban areas which provide a wider distribution of services, opportunities in agricultural production and opportunities for alternative employment. The transport system provides much more than a means of evacuating commercial produce and conveying migrants to the towns; it also makes possible modern technological change in agriculture as a routeway for information and for the supply of technical goods and services. It has been claimed in northern India, for example, that the clear correspondence between accessibility and the rate of modernization was more a reflection of information flow differences than of transport cost differences (Wilbanks, 1972).

In the development of export crop regions the evolution of modern transport and port systems has played a vital role. In the early stages of development inland transport, often by head porterage, was costly. Export crop development could affect only a narrow fringe of coastland, served by a multiplicity of small ports employing only the simplest facilities, whose distribution pattern was governed by physical geography and trade potential (Hilling, 1969). Increasing demand for tropical and subtropical produce led to the expansion inland of production, but a big advance had to await the introduction of railways and later of modern road systems. These concentrated on fewer ports serving larger areas, a tendency which was increased by the need for revenue and immigration control and by the introduction of special cargo handling facilities to deal with the increasing volume of traffic and to improve the efficiency of the service. Port and transport system evolution have been modelled by Taaffe, Morrill and Gould (1963) in a study capable of wide application and interpretation although concentrated mainly on West Africa. The export crop and related regions which were described above could not have been created without modern transport and port structures, and their further evolution must lead to further modification of such regions. The Third World countries share with others the problems of road and rail competition and the tendency for railway costs to rise faster than those of road transport. Most effort is today concentrated on developing road systems, but new railways are still being planned and built, chiefly for the movement of bulk goods, as the proposed Trangabonese Railway to exploit the timber, manganese ore and iron ore resources of Gabon testifies.

Changes in port technology have led to greater concentration on some forms of agricultural production, and urban primacy has encouraged the growth of large areas of vegetable and other fresh food production in the vicinity of huge cities. Other developments, such as the processing of agricultural produce for the internal market, perhaps surprisingly, have not always had similar effects. In Thailand, for example, the introduction of steam-driven rice mills in 1858, initially served by water transport, encouraged a marked concentration of commercial rice production. After 1918 road building favoured the decentralization of milling, and the latest petrol and kerosene fuelled mills are very small and can be established successfully in a very wide range of locations, although their tendency to produce a more broken rice has made them less suitable in the long grain growing areas of the central plain (Hafner, 1973). In Malaysia the pineapple industry was concentrated firstly by linkage to rubber cultivation and secondly by the dominance in the production system after 1945 of the canneries with their own estates. The slump in the canned pineapple market of 1956 led the cannery owners to refuse to process the fruit from smallholdings. This was followed by political action, a government enquiry and the eventual building of government canneries which widened the area of production and made it possible for smallholders to increase their share of production from less than a third in 1960 to over two-thirds in 1970 (Wee, 1970; Tay and Wee, 1973).

In the Third World assumptions of movement minimization, often based on the observation of high transport costs, are not always sound. Transport hinterlands of markets or pick-up points in relation to farmers are often very difficult or even impossible to determine because of the complexity and variability of the arrangements. Plumbe (1974) has cited the purchase of cashewnuts and copra in Tanzania by co-operative buying posts. This was complicated not only by the establishment of temporary subsidiary buying posts to lessen the burden of crop cartage to the farmer, but by a periodic lack of cash and by embezzlement amongst individual co-operatives, which resulted in the closing of societies for up to two weeks even in the peak harvest season. Farmers also tended to travel to whichever society was thought to be currently purchasing rather than delay realization of cash. This pattern of behaviour was encouraged by pick-up operators, although not at extra cost to the farmer, who paid a standard charge per unit of load regardless of distance. Pick-ups even plied in search of societies making purchases, usually moving towards Dar-es-Salaam, so that co-operatives near the city made larger purchases than those farther away. Plumbe concluded that attempts to delimit hinterlands were sterile. Many other examples of this kind could be cited. In Western Nigeria, for example, cocoa purchases at buying centres were not strictly related to the productivity of supposed hinterlands, and farmers would move to distant buying centres if they thought there was an advantage in so doing.

The infrastructure that is so essential to commercial agriculture itself creates its own demands for food and for raw materials, as do the industries which today accompany urban growth in the Third World. Many Third World cities have grown from small trading beginnings in which the handling of the produce of cultivation, gathering or mining was their main *raison d'être* and from which some of the profits have been invested in urban and industrial development. They now emerge as so-called growth poles with increasing industrial productivity and commerce, creating in their turn a profound effect on the surrounding rural area. Thus around Calcutta the rural population densities are high in response to a virtual urbanization of agriculture in which the dominant elements are market gardening, fruit production, dairying and poultry keeping. A perishable goods producing region of huge extent has formed, subdivided into a number of specialized or near-specialized sub-regions (Dutt, 1972). In Northern Nigeria the traditional social linkages between rural and urban dwellers have produced a similar zone of urbanized rural concentration around Kano, but only in part for market gardening and fruit production to feed the city, which also looks to the rural area for wood fuel and as a disposal zone for its nightsoil. Here the linkage is mainly social and political, for the rural population regards Kano as the natural hub of its existence (Mortimore, 1975). Pressures on land resources and demands for nearby land have increased as modern commercial and industrial developments have taken place, but the foundation of the current pattern is traditional, and modern developments have emphasized existing trends.

REGIONALIZATION AND AGRICULTURAL CHANGE

The regionalization of agricultural production in less developed countries is complicated by the highly varying speeds of agricultural change and by its variable character. Generally one may contrast traditionally oriented and often extensive systems of agricultural land use having low yields, low real incomes and low levels of labour utilization with more modern, usually more intensive, more specialized, higher input systems employing regular wage labour. A major problem in effecting change is the level to which the extensive sector can continue to satisfy rising demand by expanding its area of production. Thus Myrdal contrasts the widening yield gap between developed and less developed countries, which is related to the widening income gap both as cause and as effect (Myrdal, 1971, p. 93). This simple dualistic thesis of contrasting intensity has much to commend it, despite criticism of dualism in general, more especially in those locations where only small areas are available for agricultural expansion and there is no longer room for agricultural migration. In such a situation farm subdivision and fragmentation may occur, but accompanied by rising levels of intensity in production. Not all such increases in intensity, however, provide satisfactory increases in income. They may be merely the means of offsetting the most serious effects of a worsening land situation and obtained at a heavy cost in man-hours, with declining marginal rates of return (see pp. 22 and 34). Improvement in productivity is sometimes associated with reduction in production intensity, by adopting extremely extensive agricultural systems, even so-called 'shifting' agricultural methods, wherever labour and land availability together with technical and capital poverty dictate. Indeed Thünen's theory would have predicted it with its cash-cropping and livestock rearing zones on the periphery of the 'isolated state'. Thus there are parallels between the rapid inland shift of estate-organized coffee production in Brazil, or of peasant-organized cocoa farming in Ghana, and the emergence of shifting commercial agriculture in Togo, Benin Republic and central Nigeria wherever young peasant farmers have abandoned the over-crowded heavily manured and terraced uplands of their fore-fathers for bushburning and cash-cropping in small villages or dispersed hamlets in the surrounding plains. All these features are the product of the same modernization processes which elsewhere have introduced modern mixed crop and livestock farming, the use of manures and fertilizers, and scientific tree-cropping.

In the appropriate location shifting agriculture may be as much a feature of modernization as more intensive systems elsewhere. But each system meets in its own locality a different combination of the factors of production, just as so-called traditional systems survive in new guises, some more intensive and others more extensive, as contact with the more developed world increases. A simple spatial dualistic model of a rising tide of intensity will not suffice. Instead we must examine the regionalized complexity of agriculture in the Third World, bearing in mind that many of the systems we examine are far from static in form, or rather far from being in a state of equilibrium. We must seek to determine the goals towards which they are moving, even if some of them will never reach their targets before they are overtaken by a new diffusion of change in technique, market or input mix. Change may not always be in the direction of rising productivity and increasing wealth. Huge areas of the Third World are experiencing poorer productivity and reduced incomes as pressure on resources develops through increasing population, mismanagement or the transfer of resources out of agriculture.

India, for example, has regions of rural stagnation and decline which are far more extensive than its modernized farming regions with water control, fertilizers and hybrid cereals. Rural poverty has increased since 1960, whilst overall incomes have risen by so small a margin that they have virtually stagnated. This appears to have been caused not by migration to the towns, because the rural proportion of population remains high, but by the rising proportion of rural poor, especially in eastern India (Bardhan, 1970 and 1973). A different view has been expressed, however, by Minhas (1970) who notes some decline in rural poverty in India. Poverty has increased especially in the chief rice growing region, where in most states nearly two-thirds of the rural population have a standard of living below the official minimum level, and even in the Green Revolution states of Punjab and Haryana, despite increasing expenditure on education, health services and rural electrification which has helped to reduce regional disparity (Gupta, 1973). It is hard to resist the conclusion that the pressure of rising numbers in India is being felt strongly in certain poor rural regions, despite the evidence of urban poverty. This pressure of rising numbers also appears to have been responsible for the major part of the increase in agricultural production being met not by the new technology of the Green Revolution, but by regional expansion, that is by bringing more marginal land into production and also by double-cropping. Yields have declined in the older areas affected by population pressure and are lower in the new marginal land areas (Chakravarti, 1971). A region of increasing rural poverty, based mainly on subsistence rice production, has emerged, whose core area is West Bengal, with three-quarters of its rural population below the minimum level of living, with declining yields at its 'centre' because of pressure on soil resources, and decline at the fringes because of expansion on to marginal land. Innovation and improved productivity together with any prospect of increased commercialism have become virtually impossible except on the few larger and advantageously situated farms or on smallholdings close to large urban markets. Regional changes of this kind are a form of spatial diseconomy which arises from a succession or chain of internal difficulties. Innovations may lead by their introduction to changes in the regional landscape which in turn can undermine the new system of production. Increased rice production has led in many cases to spectacular overcrowding which has made it impossible to sustain high yields. Irrigation schemes have led to soil impoverishment, as have some of the methods used to extend the area under tree crops. Continuous areas under new crops have encouraged the spread of disease, and the generation of new virus forms appears to be engaged in a race with the breeders of new crop varieties and the manufacturers of chemical sprays. Equilibrium can be maintained only by a race in which the costs of the new techniques are a constant drain on the profitability of the new system. In the waves of innovation which may occur we should try to distinguish between those which arise from external forces and may raise productivity to new levels by fundamentally changing its character, and those which arise from internal forces, that is are responses to the internal difficulties arising from previous innovations and seek mainly to maintain the production *status quo*. In practice such distinction may be virtually impossible as new techniques, which may alter the character of the entire system, may be discovered, perhaps by accident, in research institutes intended to cope with current difficulties, whilst controls needed to maintain the system may well be introduced from outside. Some of the complexities of changing regional systems will be illustrated below by an example of one of the kinds of change which have most affected Third World agriculture, namely the spread of tree crops, using as a case study the largest and most spectacular spread, that of coffee in Brazil.

As agricultural development takes place its spread not only changes total production with consequent effects on the market and subsequent response from producers, but is also frequently accompanied by changes in technology and the supply of inputs which promote cultivation in some areas but discourage it in others hitherto successful. Development for some regions may mean a brief period of prosperity followed by worse poverty than existed before. There are several examples of this in the West Indies, where islands provide discrete units of production, many of which are today too small to maintain certain services. The island of Nevis, for example, is increasingly dependent on subsistence cultivation, and sheep and goats graze on abandoned estate land as a result of development. Absentee landlordism, decrease in capital availability, difficulties in trade and shipping, some soil erosion, the high costs of milling sugar cane from a small producing unit and rising costs of ferrying to the neighbouring island of St. Kitts have all produced a decline in cane production (D. Watts, 1973). One of the richest islands in the Caribbean has now become one of the poorest. This could be a pointer to future decline in many other regions of the Third World if the pace of technology and yield improvement becomes much faster than the capacity of markets to absorb the productivity gained, so that the more successful regions deprive others of their markets.

THE COFFEE REGION OF BRAZIL

Coffee was first introduced to the Pará region of northern Brazil in 1727 at a period when sugar production based on Indian slave labour was declining both in Pará and in neighbouring Maranhão. For a time it succeeded, attracting capital and labour from the sugar growing region of the Northeast, where between 1780 and 1790 some 150 000 people left for the coffee frontier of Maranhão (James, 1959, p. 418). A few successful attempts were made to establish coffee plantations in the Northeast, but the alternatives of emigration to the diamond fields of Minas Gerais and southern Baía, or reorganization and re-equipment of the sugar estates, generally proved superior, and in the uncertain climatic conditions of much of the interior Northeast cotton proved a better new boom crop, with cocoa as an alternative in the wetter coastlands of southern Baía.

In 1774 coffee was introduced into the hinterland of Rio de Janeiro and the Paraíba Valley, again largely on former sugar plantations. By the late eighteenth century Brazilian coffee was established in the European market, demand was rising and the new crop expanded rapidly. For the first time coffee became pre-eminent in the Brazilian economy and the plant was displayed with tobacco in the coat of arms of the new Empire of Brazil in 1882 (Sternberg, 1955). Competition from the northern plantations disappeared as northern labour was diverted almost universally to rubber gathering and transport, which lasted until the collapse of 1910–12. The main advantages of the Paraíba Valley were cheap land and proximity to the great port of Rio de Janeiro, but poor soils and steep slopes led to early decline in production and rapid advance inland to southern Minas Gerais, where yields were low and plantations soon abandoned, and then to São Paulo. By 1875 the Central Railway had been extended to the coffee planting district of Juiz de Fora in southern Minas, but only two years later the branch line reached São Paulo city, too late to save coffee production in the Paraíba Valley, but at just the right time to promote a great coffee boom on the small area of terra roxa soils on the lava beds of São Paulo. This was followed by planting on the much more extensive lava outcrops of Paraná, occupying some 60 per cent of the area (James, 1959, p. 489 and Sternberg, 1955). Rapid advance was always encouraged by the cheapness of new land and the speed of railway and road construction. From 1850 onwards the pace of immigration increased. 4·6 million immigrants entered Brazil between 1884 and 1954—32 per cent Italian, 30 per cent Portuguese, nearly 14 per cent Spanish and 4 per cent German. In the late nineteenth century most of these went to work on coffee estates or established their own farms. They were especially attracted to São Paulo State, where the *colono* system of small farms, usually worked by immigrants on large estates, was first established (at Limeira in 1847).

As early as 1882 the State Government helped to break up large estates and to establish immigrants on their own farms either by direct sales or through the agency of private land companies approved by Government. São Paulo city became the chief focus of development, partly because of its function as state capital, its existing importance (a city since 1711, possessing one of Brazil's two academies of law and a population of 35 000 by 1879) and its position as the centre of the state railway system. Coffee planting spread northwards on the crystalline plateaux sharply limited by the lowlands to the west. But it also spread across the lowlands to the northwest on to the diabase plateaux especially around São Carlos and Ribeirão Preto (Fig. 24), where the terra roxa soils gave the highest yields. The chief focus of this initial spread to the north and northwest was not São Paulo city but Campinas to the north, which itself had a small area of terra roxa soils on a narrow diabase dike on which stands the Instituto Agronómico de São Paulo, the greatest of the world's coffee research centres. From 1860 to 1885 Campinas was the chief centre of coffee marketing and for a time grew faster than São Paulo city as planting expanded most rapidly on the uplands served by the Paulista Railway. From 1885, however, planting spread westwards into the huge area served by the Sorocabana Railway system, focusing on São Paulo city. Much more extensive coffee planting was undertaken on the huge diabase cuestas extending westwards and northwestwards from Botucatu, and across the state border into Paraná. São Paulo city grew rapidly, becoming the main centre of the coffee trade and the main centre of investment of coffee profits. This development took place partly because it was the state capital and an established city, partly because it was the last 'bottle-neck' on the routes to Santos, for a long time the dominant exporting port, partly because it was higher and cooler than Campinas and less subject to fever epidemics and partly because of accessibility to cheap power (Cubatão has been claimed as the finest water power site in South America). São Paulo also controlled cotton marketing and dominated cotton textile production, which by the 1940s employed nearly 40 per cent of its industrial workers. Political control was an important factor in maintaining the pre-eminence of the coffee industry and the ascendancy of São Paulo, chiefly through valorization policies adopted by a government largely controlled by the coffee growers until the economic collapse and defeat of the coffee party in 1930 (James, 1932). The industry up to 1930, despite attempts to break up some of the estates, was still dominated by tenancy arrangements under the *colono* system. Much of the wealth went to landlords resident in the towns, more especially in São Paulo, where even in the early 1960s the Avenida Paulista still contained the villas of the coffee barons. Japanese, Italian and Jewish quarters marked successive 'layers' of urban settlement, matching in a sense the 'layers' of rural settlement which marked each stage of immigration.

The expansion of coffee planting in the 1920s was encouraged not only by boom economic conditions until 1930, but by the increasing tendency for frontier coffee farming to be developed less by large estates and more by independent small farmers. Thousands of former tenants emigrated to a pioneer existence on the frontier in the hope of achieving independence. It was also encouraged by the increased rate of decline of the older plantings, especially on the poorer soils, due mainly to pests and disease, in particular the attacks of a beetle called the Coffee Berry Borer (*Stephanoderes Lampei*). The result was a patchiness in the coffee distribution in the older planted areas, increased attention to other crops and intensification of production, i.e. the use of insecticides, fertilizers and improved plant stock. Whilst some plantations were thus improved others were abandoned as labour left the area for either the towns or the new coffee frontier. Figure 24 shows the distribution of coffee production in 1927–8, indicating areas of growth and decline since 1910 (Morgan, 1973, p. 17). Some diversification in agricultural production has long been characteristic of the coffee growing area, partly due to the abundance of land generally unsuitable for coffee, the demands of both agricultural and urban workers for foodstuffs and the readiness of farmers to plant other boom crops or to seek some insurance against commercial risk. Cotton has already been cited and was adopted by growers more especially on the western frontier, i.e. the traditional coffee frontier, in the 1920s and 1930s as industrial demands rose. Occasionally cotton even replaced coffee, but mostly it replaced associated food crops, maize, rice and beans, grown usually in locations less favourable for coffee such as valley bottoms and slopes. Cotton was followed in the late 1930s by oranges, especially in the Paraíba Valley and the Paulista Railway zone, and more recently by soya beans, the latest boom crop, encouraged by the trend to mechanization as labour costs rise, an expanding home market and exemption from export taxes (Burley, 1973). Yet despite this tendency to diversification coffee tree plantings continued to increase and during the 1950s production rose faster than ever. Most of this was achieved by further expansion into new lands despite increasing distance from São Paulo and Santos and despite increased environmental hazards. A 'hollow frontier' developed as old areas were abandoned, mainly to new crops. São Paulo State has become a highly diverse agricultural area with about a third of Brazil's coffee production, 70 per cent of its cotton, half the sugar and a third of the rice. The leading coffee state in most years became Paraná, with over half total production by 1969–70 (severely reduced in 1970–1 in total and proportionately less than São Paulo in 1972–3). In consequence Paranaguá 167 miles to the south has taken the place of Santos as Brazil's leading coffee-exporting port. The 300 mile two-lane highway from Paranaguá to Londrina is today the chief coffee export route.

The expansion westwards and southwards into Paraná had the advantage of considerable extents of *terra roxa* land superior to the worn soils of São Paulo or to the coarser, more leached soils to the north where high temperatures (associated with increased disease risk, especially from coffee leaf rust) proved inhibiting. To the northwest the chief constraint was aridity. Paraná with its frosts was the preferred risk because of superior soils and accessibility to port, helped by state government encouragement of colonization schemes. Coffee planting has even crossed the international frontier into Paraguay where Brazilian investors have bought land near the Brazilian borders.

In Paraná killing frosts occur in July and their frequency has proved remarkable in terms of periodicity and possible prediction (Taylor, 1974). Frost damage has been so severe that coffee production has become highly variable, with yield reductions of 50 per cent or more. Between 1953 and 1971 inclusive there were five severe frost years. On 5th July 1953 220 million coffee trees were estimated to have been burned by frost, which was a direct cause of the Brazilian economic crisis of 1954 when coffee production still remained low. Another severe frost followed in 1955 and a major although less severe frost in 1957. Subsequently production boomed, although with somewhat variable results, until another severe frost in 1963, but this had the result, in a period of overproduction, of raising price levels and, for many growers, of increasing profitability. In some cases also frost damage has helped to encourage the exchange of coffee stocks to support Brazil's rapidly growing instant coffee industry. In 1969 a very severe frost in northern Paraná damaged 97 per cent of the crop, although other areas were less affected, and for the coffee region as a whole this was a major rather than a severe attack. Finally in 1971 yet another severe frost was associated with the lowest production for nearly 25 years.

With the move across the frontiers and the severity of the environmental constraints easy and rapid expansion of coffee production seems finally at an end. Labour costs in coffee production are high and mounting. Increasingly the tendency is to restrict planting to the better soils, less hazardous and more accessible locations, to use superior stock, fungicide sprays and irrigation (as a protection against frost), to introduce tractor cultivation with wider rows of individually planted trees instead of the old four-in-hole system. In an area of overproduction with little virgin land left intensification makes economic sense, as does the substitution of capital inputs for labour if greater stability in production and in the world coffee market can be achieved. The threat of a reduced labour force is, however, serious because it promises lower levels of rural employment unless the discovery of a frost-resistant coffee strain opens up a new era of expansion or alternative industries can be developed. Many of these people will move to the towns, more especially to the slums of São Paulo, already overcrowded with the *flagelados*, the peasant victims of droughts in Baía and Minas Gerais.

traditional prices. 'In the field of economic relations the population of peasant producers is characterized by its heterogeneity, the small population, the relative independence of each producer, and the limited output of each farmer. It is heterogeneous in the sense that not all producers share the same aspirations, the same knowledge, or the same responsibilities or hold the same amount of capital assets; each unit of production is different from the next' (Ortiz, 1973, pp. 1–2).

Some small farms have a spatial patterning within them which affects their productivity and which may be studied as a micro-aspect of agricultural geography. Others, extremely small and compact, have little spatial patterning within them worthy of note. The general lack of medium-sized farms, say approximately 20–200 hectares, in the Third World, is an important feature of the spatial organization of production, and a key to several development problems. Much emphasis in recent years has been put on the need for large-sized production units to achieve economies of scale, and, whilst such scale economies are important in some contexts, it is often not appreciated that the very large production systems so characteristic of agricultural production in certain locations in the Third World are often a reflection of a peripheral development situation in which the success of the system depends on low inputs or on the low incomes of the larger part of the workforce. Both very large and very small farms have been equally characteristic of low productivity situations and of economic and even political dependence on wealthier nations.

The case for 'micro-research' in the geography of development, more especially at the village level, has been forcibly argued by Connell (1973), who believes that much needed micro-empiricism must precede most macro-theory and that the most profitable geographical insights, both practical and theoretical, will come from work done in rural areas, especially at a small scale. Many of our scientific enquiries have to begin at this scale, which is also the most important decision-making level since, whatever governments may decide and environments constrain, the final choice of enterprise must rest with the farmer himself. Although we write of 'scale' and of 'micro-research', size is not in itself an absolute criterion. The size range between small farms and farms affiliated into village systems or estates is considerable and overlaps the regional and sub-regional scales already discussed. Some plantation firms have enormous holdings scattered in several countries in order by geographical dispersal to spread both environmental and political risks. Such units are of an international character and can hardly be considered in the same context as peasant farms of less than a hectare, yet in both we are concerned with the spatial organization of production systems. Brookfield (1973 and 1975, p. 169) has argued that the real meaning of the macro/micro distinction is not one of scale, but rather of aggregation or disaggregation, instancing the economy of the Republic of Nauru as a rather extreme case of macro-study, and the structure and operation of Rio Tinto Zinc as an extreme case of micro-study. There are many instances of serious size anomalies; nevertheless, as Brookfield admits, most examples correspond with the meaning of the two terms, macro- and micro-study. Undoubtedly the general scale approach has practical advantages.

The main factors in the spatial organization of farms or of groups of farms are the inputs, more especially the labour and the quality and quantity of land available. The accessibility of the fields to the farmers and their labourers may be seen in many instances to be the key factor in the spatial patterning of enterprises and of agricultural techniques, but often the spatial arrangement of the farmers themselves is determined mainly by non-agricultural considerations, and so other factors somewhat external to agriculture also operate as major determinants of such spatial organization. Relationships to land in small farm systems are clearly very different where farms are dispersed from the situation where farms are closely grouped together in large villages. The larger the village the greater the distance to farthest fields and the greater the time spent in the journey to work. The social system governs the allocation of tasks between the sexes and age groups, relationships between farmers, villages and towns, marketing, choice of enterprise and even the alignment of the mounds and ridges created by tillage. It also governs land tenure, inheritance and changes of location or of property with marriage. Often air photography indicating patterns of agricultural activity is as useful a guide to social condition as it is to the spatial ordering of agriculture. The relationship between society and farming has been particularly emphasized in shifting cultivation, where de Schlippe has seen traditional agriculture as almost synonymous with 'the culture as a whole in its function of ensuring the survival of the group in its habitat' (Schlippe, 1956, p. 241), and Conklin has stressed the cultural context in his 'ethno-ecological' approach (Conklin, 1954; Watters, 1960). Compact settlement may not only reflect village organization as an aspect of local social system, but may be the product of modern socio-economic planning aimed at creating organizations thought more suitable for the introduction and development of new techniques of production or offering superior social incentives. The *ujamaa* villages and the development of co-operative farming systems cited above (pp. 140–1) are cases in point. In a very different socio-economic context planned estate villages, grouping tenant farmers in a common farm scheme or grouping agricultural labourers on a plantation or in a state farm, serve a not dissimilar purpose and achieve similar spatial results. A plantation may achieve something of the spatial form of a market-centred system where its focal point for agricultural operations is a processing mill acting in a similar spatial role to a market in relation to the product. Generally, however, processing marks another stage in the production process and lies outside farms. Inputs also include capital, which usually has little or no spatial effect on the farm itself. There is, however, a tendency in some farm systems to spend more on fixed and working capital, to invest more in the nearer and more accessible fields. This reflects in part management convenience. Such fields can be visited more frequently and receive more labour and more attention from management. They can also be more easily protected from stealing and from pests. Often tree crops of high value are located adjacent to villages or next to the farms themselves, but there are many exceptions.

The enormous ethnic variety of Third World countries is reflected in considerable variation in social condition, which in turn is reflected in variation in the agricultural space-economy. Thus amongst the Hausa of Northern Nigeria social and political relationships have created hierarchies of settlement in which many or indeed most of the farm villages are dependent on the towns as markets, as sources of manure and manufactured goods and as social centres. Settlement organization and a reluctance to farm more than 3 or 4 miles from home have produced an immense crowding of people around a few major foci. Non-agricultural activities, especially trading, are closely linked to agriculture. Frequently the larger holdings have been acquired by men who have become rich through trading, whilst a small additional income from trading is essential for farmers whose holdings are too small to provide an income sufficient for subsistence. But inheritance laws and other difficulties stand in the way of amassing capital and impede the development of a wealthy farmer élite, as in some regions of India, despite increasing income disparity between rich and poor farmers (Hill, 1972). Renting has become more common, not only for poor farmers seeking larger holdings, but for richer farmers seeking expansion. Tenancy in less developed countries has often been condemned because of landlord's exploitation, but in this case tenancy is not necessarily detrimental to development. In Kelantan in West Malaysia the Muslim laws of inheritance combined with local crowding have encouraged excessive fragmentation and uneconomically small plots. The more commercially oriented farmers rent additional land and have become what Huang (1975) has termed 'owner-tenants' who tend to have larger farms than 'owner-cultivators'. By contrast, in Guyana the East Indian farm population has suffered much less than the neighbouring Negro farm population from fragmentation and reduced holding size. East Indian farmers encourage their sons to farm and to acquire land, but usually near the family home. Family resources are pooled and invested in more land, which is regarded as a form of security, and in machinery (Burrough, 1973). Here a different social tradition has encouraged the more successful farmers to choose not rapid expansion by renting but a slow expansion by land purchase with a view to long-term benefits for the family. In part this may be associated amongst Hindu East Indians with an ethic which emphasizes prestige in the local community. The result is smaller, more compact and more permanent farming units than those established by the more commercial and successful Hausa and Kelantan farmers providing a contrast in the farm space-economy. Muslim East Indians are less separated from Hindus than elsewhere and share many social traditions and features encouraged by a common attachment to land, patrilineal inheritance (which discourages fragmentation), availability of land for purchase after completing indenture, and fear of Christian indoctrination in urban schools (Lemon, 1975). The examples of contrasting spatial organization can be multiplied elsewhere. Especially notable have been Uhlig's attempts to develop a systematic social and ecological differentiation of agricultural landscape types summarised in his study of hill tribes and rice farmers in the Himalayas and Southeast Asia (Uhlig, 1969).

The farmer must choose his mix of enterprises and in some cases from time to time may choose the location of part or all of his holding. Both decisions are locational. The locational aspects of the individual farmer's choice of enterprise mix have been largely discussed at other scales because only there is the pattern effect observable, and the question of individual choice, although interesting in itself, is much more interesting geographically where a large number of individuals is involved. The question has important implications for advisory, extension and research workers who may seek to solve regional or 'industry' problems, but have to work at the farm-firm level, where the interests of the individual and the appreciation of the factors involved may differ considerably from the interests and appreciation of the group. In so far, however, as choice depends on the spatial patterning of agricultural activity, as in the spatial requirements of a rotation system or the spatial relations of a mixed system of livestock and crop raising, some aspects remain to be examined at this scale, together with the related question of the location of all or part of a holding. There are not many migratory farmers or 'expanding farmers' in the Third World who can make a real choice of location to develop their farming activities. Most such deliberate choice of optimal or sub-optimal location of farm production has been made by large, usually expatriate companies seeking suitable sites for highly specialized export crop production. Successful small farmers able to acquire additional land to expand their holding would be very lucky to acquire adjacent land in order to keep their operations compact. They tend to buy land as they have capital available and as land is offered for sale or rent. They may have some choice of land quality and location, but it is usually very limited, which is why laissez-faire expansion is not the solution where it is desired to increase the size of the blocks of land farmed in order to permit mechanization. We may re-order the questions raised from the original two general location questions into three groups: the spatial structure of farming systems, accessibility and zonation of land-uses, and parcellization or the location of areal units of production. For convenience the first of these will be considered with regard to small farms, and then with regard to estates, plantations and other large farms. Zonation will be considered as an aspect of the spatial ordering mainly of small farms, but separate consideration will be given in conclusion to parcellization as a feature of spatial discontinuity.

THE SPATIAL STRUCTURE OF FARMING SYSTEMS: 1. SMALL FARMS

The varied husbandry techniques have each its own spatial expression or patterns of crop and pasture location in relation to farm buildings, access to other members of the community and access to market. These patterns are in themselves a constraint, restricting choice of technique to areas where they may be developed, and forcing technique substitution where other constraints prevent their development. The variety of techniques and spatial patterns is too considerable for thorough review, but some general notions illustrated by a few examples can be indicated here.

a. *Shifting cultivation.* The term shifting cultivation, which is so widely used in the literature on Third World agriculture, refers to a wide range of techniques, mainly occurring on small farms, although not absent from large, in which there is a rotation of fields rather than of crops and in which short periods of cropping (usually one to three years) alternate with long periods of self-sown fallow (normally six to rather more than twenty years) (Watters, 1960). The brief rotation of crops as such within a field is often irregular and has been described by the term 'pseudo-rotation' (Schlippe, 1956). Other terms include the archaic 'swidden' (Izikowitz, 1951; Conklin, 1954) and a great variety of regional names, such as *ladang* and *conuco* (see pp. 65–8). Crop production on any particular piece of land is largely governed by length of fallow, which in turn depends on population density and distribution, the market and soil and moisture conditions. Only low densities of population can be supported by shifting cultivation, frequently even less than 12 per square kilometre. Where shifting cultivation is combined with other techniques in a system of varied farming methods it normally has a peripheral location and is often associated with crops needing little attention. It is doubtful whether in shifting cultivation the fallows represent a conscious attempt to restore fertility rather than abandonment of land which no longer gives satisfactory yields. Often a return to a site formerly cultivated results in the creation of new field boundaries, so that there is no regular rhythm in the use of particular sites. The burning of cleared materials that accompanies shifting cultivation does not normally appear to be allowed to spread. Pulled grass and the remains of herbs, shrubs and trees are heaped and allowed to dry before being set alight. Savanna fires appear to be associated mainly with hunting, accidental burning or wanton acts of destruction. Settlement is normally dispersed where shifting cultivation is the sole or dominant farming technique, as large villages incur problems of access to farmland which are most easily solved by the application of more intensive techniques to the nearer fields. Settlement migration is not necessarily involved and where it occurs may be more for sanitary or social reasons than because of locally declining soil fertility. Not all crops give poor yields after only one to three years of cultivation. Some, such as opium, give satisfactory yields for periods of even up to twenty years of continuous cropping, and in consequence have been highly valued wherever the tasks of clearance have proved especially heavy or where land has been scarce (Geddes, 1973). Shifting cultivation is not restricted as a technique mainly to subsistence farmers and has frequently been employed to produce commercial crops wherever land is plentiful.

One of the most thorough studies of regional variations in shifting agriculture technique with attempts to assess their carrying capacity is Allan's *The African Husbandman* (1965), which compared a number of practices, mainly in Central and East Africa, including a special study of the *citemene* methods of cultivation in Zambia. *Citemene* cultivation supported fewer than 4 persons per square kilometre with woodland regeneration periods of over 20 years and cropped areas at any one time of 0·1 to 0·4 hectares per head. An important feature of these practices was the cutting of timber over a wide area in order to provide very large quantities of ash for the small area cultivated. As much as 2·5 hectares of *miombo* forest (*Julbernardia—Brachystegia* woodland) were required to provide sufficient ash for 1 hectare of eleusine millet. Allan distinguished two broad classes of *citemene*. The first, seemingly more primitive, consisted of the 'small-circle' methods, that is several circles of land some twenty to thirty feet in diameter were established in a large cleared area and used for stacking tree trunks and branches for burning. Eleusine seed was sown broadcast in the ash patches and most of these gardens were abandoned after only one year. In 'large-circle' methods the woodland was only lopped, not felled and all branches from one cleared area were stacked in one circle about 0·4 hectares in area, and burned and the area planted to eleusine, but used subsequently for a sequence of crops grown in rotation. On the whole 'large-circle' methods seemed able to support slightly more people per square mile than 'small-circle' methods, but involved more work in lopping trees and producing enough ash on which to base four or five years of rotation of such crops as sorghum, maize, cassava, beans and groundnuts or a second planting of eleusine, although the crops grown after the first year were not usually broadcast. All the cultivators had subsidiary hoe gardens, many of which were sited on or near *dambo* or floodland, where more intensive methods of production were based on tillage instead of sowing broadcast in ash. There is evidence that the fullest development of *citemene* cultivation was an early result of colonialism. In pre-colonial times men were admired for their skills in hunting, tree cutting and warfare, but not in agriculture (Richards, 1958). In the early stages of colonial rule, local warfare ceased and the population became dependent on subsistence cultivation with plenty of male labour available for farming. Subsequently the men began to move to the mines in search of work, reducing the labour supply to farming which increasingly was left to women, children and old men. The lack of the labour needed to cut or pollard trees has reduced the importance of *citemene* and increased dependence on hoe gardens, where cassava and maize have acquired much greater importance. *Citemene* was also discouraged by the administration, which thought it destructive of vegetation and believed the movement of villages to be inimical to the improvement of social services (Richards, 1958 and Kay, 1967, p. 56). Improved farming schemes in the 1950s were intended, as Richards reported, to 'stabilize the population on the better soils'. Thus the spatial structure of Bemba farming is changing both in form and in mobility as settlement becomes fixed and agriculture becomes more intensive.

Most shifting agriculture uses the timber and other vegetation of an area not much larger than the planted field, which is often irregular in shape. In general shifting practice is restricted to rainforest or well developed savanna woodland. In the rainforest, burning is often limited by the lack of a dry season and consequent difficulties in drying timber. In Ghana Manshard (1957) has described the '*proka*' technique of clearance without burning, but it appears to take place mainly in the wetter areas and consists of removal of the undergrowth in order to plant cocoyam and plantains. Where population density increases shifting agriculture may continue at first with shorter fallows, but eventually shifting as such must cease and other forms of agricultural practice take its place. The relationship between population density and shifting agriculture has already been discussed (pp. 30–4) and has been described in relation to specific instances in a considerable literature. It is an important relationship to consider for land-use planning, but it is complicated, as Datoo (1973) has shown in Tanzania, by the very large number of variables involved. Datoo's techniques of statistical analysis and multivariate classification offer a possible solution to the problem of discerning the relationship, but the results even with these refinements of technique are not easy to interpret. Few examples of husbandry practice fall neatly into pre-determined classes or suggest from any intensive survey the clear breaks in ordering the data the typologist may seek.

Shifting agriculture does not necessarily disappear in the face of a changing economic situation. It has its advantages for low cost production on cheap land with little or no capital and may persist, although commercialism inevitably demands some degree of settlement and land use concentration. In the former Belgian Congo the attempt was made by the Institut National pour l'Etude Agronomique du Congo Belge (INEAC) to rationalize shifting cultivation by developing improved farms or *paysannats*. In rainforest the 'corridor system' was introduced in which the areas cultivated were laid off in long parallel strips each 100 metres wide and aligned approximately east to west. The width was calculated to be wide enough to minimize over-shading and border effects and narrow enough to encourage fast regrowth of fallow. Regular rotations and fallow periods were maintained and land allocation followed local custom, allowing each family to cultivate as much as it wished. In savanna, block lay-outs were used, subdivided into individual family holdings or *fermettes*, again with regular rotations and lengths of fallow. Fixed areas were allocated to each family in the savanna system, which was more a system of rotational bush fallow although with long fallows of thirteen years or more. The fixed area had disadvantages in relation to the variable labour performance of different families, but was governed by the need to control the area cultivated in order to relate it to the provision of mechanical equipment and services (Allan, 1965, pp. 437–45).

b. *Rotational bush fallow.* Some research workers include all techniques which depend on clearance and burning of self-sown vegetation in one single class of 'shifting cultivation' or 'land rotation'. Others, however, follow Faulkner and Mackie and recognize a more intensive group of agricultural practices in which 'the time in fallow exceeds the time that the land is cultivated' (Faulkner and Mackie, 1933, p. 44) but in which population densities are higher and all 'high forest' has disappeared. The type is similar to Allan's 'recurrent cultivation' or 'land rotation cultivation', in which the relationship between land and population is more stable (Allan, 1965, pp. 30 and 33–4). Most rotational bush fallow practice is aimed at creating a regular succession of self-sown fallows and may even include the planting of small useful shrubs in the fallows. Fields are usually rectangular and can be identified after several cycles. Possession of certain fields may also extend over several cycles although such possession may be in terms of a lineage or extended family rather than of an individual. Population densities may in extreme cases be as high as 230 persons per square kilometre, but mostly are in the 40–120 range. Fields are usually in large blocks, extending over a considerable area and including patches of fallow and cultivation. The edge of any maturely developed woodland or forest is sharp and distinctive, for woodland and forest are rarely incorporated into any agricultural system in which the fallows are regularly rotated, and usually occupy land set apart for special purposes, too poor for cultivation so possibly acting as a boundary zone between rival communities. Where population densities are high, problems of fertility maintenance can be acute, and there are many examples of cultivators applying animal manures, burying crop refuse or using techniques such as crop mixture, rotation and succession planting in order to raise overall levels of productivity. These have had some discussion above (pp. 31–2 and 62). Essentially crop mixtures allow a high density of crop plants, usually with varying demands on soil nutrients, reducing the task of weeding, and taking advantage of the small variations in drainage and soil depth that result from tillage mounds or ridges. The members of the mixture are normally planted not together but in succession, spreading the seasonal labour demand and permitting some plants to become well established before others compete for nutrients. The whole complex can then be succeeded by another complex in the second year, which may include some plants, usually minor members of the mixture from the first year. This may then be succeeded by several others, allowing a rotation, usually somewhat irregular, over possibly several years. In West Africa, in areas with a long dry season, rotations have sometimes been observed to last for longer periods than in areas where dry seasons are short. The dry season provides an important annual rest period, often accompanied by the grazing of livestock on crop refuse, which consequently allows a better recovery of soil nutrient status each year.

c. *Permanent and nearly permanent cultivation.* Wherever fallows are non-existent or the proportion of land in fallow is very small, agricultural practice may be described as permanent or nearly permanent. Such practice may occur on extremely fertile soils, such as tropical soils weathered from basic volcanic rocks. It most often occurs, however, where population densities are high enough to produce a situation of agricultural overcrowding or where problems of access to fields result in an especially high value being put on nearby plots which can be visited frequently and which may therefore repay more intensive methods of cultivation. What has been called 'compound' or kitchen garden land, often heavily fertilized and close to the settlement, is frequently subject to permanent cultivation in order to produce a great variety of vegetable crops sometimes in association with tree crops. Wherever population densities of 250 or more persons per square kilometre occur, fallows virtually disappear and are replaced by areas of continuous cultivation in small fields on small and often fragmented holdings, where cultivation frequently extends on to poor soils and up steep slopes, sometimes terraced. In such situations crop rotations may become regular and consist of repetitions or sequences of crop mixtures which can be recognized as rotation types. Some manuring systems are quite sophisticated and even involve stall-fed livestock and the carrying of manure out to fields. In some areas pressure of demand on timber resources, especially for firewood, has so reduced them that manure has to be dried and used as fuel instead. In areas of very small holdings and low incomes farmers may be forced into dependence mainly or even almost entirely on subsistence cultivation rather than incur the risks of the market, although in most years crop specialization and dependence on the market might have given a better return.

d. *Floodland cultivation.* Floodland cultivation occurs wherever more intensive and stable systems of production are preferred and where regularly flooded land is available. It may be preferred to systems of controlled water supply or irrigation wherever low inputs rather than high inputs and the maximization of output look the more profitable alternative, and wherever irrigation requires large capital works and capital is either not available or is required for other developments. The practice of planting on floodland as water levels decline is widespread in the Third World and encourages the concentration of farming in broad, gently sloping river valleys subject to annual flooding, usually associated with rain-dependent cropping on somewhat larger areas of terrace land. The mixture of practices that can occur on a single holding results often in an elaborate spatial organization with rice fields in the valley bottoms, root crops and maize on the lower slopes and fruit trees on the upper slopes. Some additional irrigation may occur, wherever ponds can be created, to water vegetable plots and sometimes to allow fish farming (Ruthenberg, 1971, pp. 142–3). Floodland normally provides higher yields, especially in the semi-arid land of the tropics, and sub-tropics, where it is most often cultivated. Sometimes mixed cropping occurs, but often planting is restricted to a few especially favoured crops, particularly rice, sorghum, maize and wheat, varieties of which may be planted in succession down-slope as water levels decline. The tasks of tillage, planting, weeding and harvesting occur not only at different times from those required by the rain-dependent fields, but are spread throughout the period of declining flood levels. Without water control the risks of crop damage and loss can be high, particularly with varying levels at the beginning of the season or at flood peak or with unusually rapid falls in flood level later on in the season. Nomadic pastoralists may compete for floodland in order to water and graze livestock, and floodland fields are often the subject of irrigation. They are also often the target for innovation, particularly by those interested in promoting water control techniques for the cultivation of high yielding hybrid grain or in promoting the use of tractors or animal draught power on land normally requiring little clearance and lacking stones or tree roots to impede ploughing. In some instances the fields have low earth banks, providing a form of 'basin irrigation', improving water retention and lengthening the period of cultivation.

e. *Irrigation.* Irrigation techniques at their simplest involve empoldering fields and constructing primitive sluices to permit water to enter at the required times and to allow flushing. More sophisticated techniques including dam construction, which can be of the simple earth variety faced with clay, or a major feat of engineering, pumping, often directly from a river or lake onto neighbouring terrace lands, and the use of wells or boreholes, either artesian or pumped. Elaborate canal systems may be constructed and the water distributed to fields by pipes, furrows or sprinkler systems. Farms may be individually operated but such systems need co-operation for effective use. In Java, for example, traditional *sawah* cultivation is supplied with water controlled by a co-operative organization, the *soebak*. A great deal of pumping in the Third World is done by manual labour, or water is simply lifted in buckets or by a *shaduf* or more elaborate Archimedean screw or Persian wheel. Animal power is frequently used, but involves the development of local pastures or the cultivation of forage crops. Generally the more elaborate and expensive techniques are associated with very high value crops grown in more sophisticated agricultural schemes, often fostered by Government. Difficulties of water and soil management are common, especially with the accumulation of mineral salts, although where it is possible the application from time to time of brackish water has the advantage of suppressing weeds at very little labour cost. Increasing problems of land shortage and of famine caused by rainfall irregularity, combined with the need for water control in order to obtain the best results in the tropics from hybrid grains and fertilizers, have encouraged the expansion of irrigation and particularly the development of great capital works. The largest single concentration is in Pakistan, although much of Egypt's once 'natural' floodland is now subject to water management control and is dependent for renewal of nutrient status not on an annual silt supply but on artificial fertilizers. Problems of disease control are also often acute, as so much disease in the tropical world is either water borne or carried by insects whose distribution is closely related to the incidence of water. The effectiveness of irrigation varies even within the small area of a peasant farm. In consequence it imposes its own spatial constraints on the farmer's crop preferences. For instance the chances are poor of water being available in the lower parts of a distribution system where water has difficulty in maintaining a high enough volume and level to flow through the ditches. Higher inputs and sugar cane were concentrated on the best irrigated sites in some north Indian villages, whilst *bajra* (*Pennisetum typhoideum*) was hardly ever irrigated (Blaikie, 1971, p. I). Where irrigation was involved minimization of movement of labour to fields, elsewhere significant as a location factor, seemed quite unimportant.

f. *Pastoralism and mixed farming*. Livestock rearing in the Third World still concentrates mainly on unimproved pasture, often rough grazings, or on the crop refuse remaining after harvest, and there have been few attempts to sow pastures or cultivate forage crops in these situations of low productivity, where a peasant farmer might consider it madness to devote human labour to providing food for animals. Nomadism is still widespread despite the introduction of ranching schemes and attempts to encourage settlement. The true nomads do not move strictly from pasture to pasture but change direction over an open range as the seasons advance, sometimes sending a herder ahead in order to set fire to old coarse grass and encourage fresh growth. Some fixed points occur as described above (pp. 194–5), including a base village and markets through which nomads have traditionally sought access to vegetable foods, especially grain. There are many examples of integration between livestock rearing and crop cultivation, although few which may properly be described as 'mixed farming' as that term is understood in the temperate world. Regulated ley farming is rare in the tropics (Ruthenberg, 1971, pp. 83–6), but it does occur in a few traditional cases such as the Sérère of Senegal with their techniques of alternate grazing and cropping, rotating pennisetum millet, groundnuts and pasture, and in upland areas, such as Ethiopia, where satisfactory grass pastures are easier to establish. Leys are becoming a more important feature of introduced techniques wherever ploughing or manure based methods have been encouraged, including some sugar estates in Cuba, coffee farms in Brazil, tobacco-ley systems in Rhodesia and modern smallholder leys in Kenya and Northern Nigeria. They provide for a more stable agriculture, often by the development of a more extensive overall use of land and sometimes involve short-term livestock movements in search of extra pasture.

g. *Mixed cultivation practice and land-use zonation.* Although there are many instances of pure shifting cultivation practice, instances of pure rotational bush fallow or permanent cultivation on rain dependent land are fewer. More intensive practices seem normally to be accompanied by a willingness to vary methods in relation to soils, drainage and accessibility, a willingness often encouraged by the use of limited nutrient resources in the form of manures, night soil, composts, crop refuse and green manures or by controlled grazing of livestock in order to apply animal manure to the crop lands. In a dispersed settlement distribution with compact holdings the amount of environmental variation within each holding is likely to be small and accessibility problems are often non-existent. The likelihood of pure practice in such a situation is high. Generally, however, more intensive methods are associated with some degree of pressure on land resources, even overcrowding, either through population increase or through spatial concentration of a population for defence, for environmental preference or for the need to be near market or a route to market. Such overcrowding is frequently accompanied by other changes, especially of land tenure and holding distribution, often resulting in fragmentation and in competition for preferred sites. In such a situation accessibility problems can occur even where settlement is dispersed. The introduction of commercial cropping may also encourage the development of different practice in distinct parts of the farm, as when the farmer finds the crop difficult to combine in rotation with staple food crops and creates separate crop combinations and rotations with the commercial and the staple crops. Sometimes problems of supervision in order to prevent stealing or attack by pests, especially birds, result in the planting of more sensitive crops near the farm and others farther away. Fruit trees, field crops such as cotton and tobacco, and some grains particularly liable to bird damage may be planted near the farm, or in extreme cases shelters are built to house bird-scarers for a short period.

Large villages incur problems of accessibility, which frequently appear to over-ride other factors such as soil differences and which have encouraged land-use zonation. Where villages have populations of 5000 or over, distances of 5 miles or more to the farthest fields have to be regularly traversed by some of the farmers. Beyond this distance farmers may find it worth while to set up a temporary shelter or seasonal homestead (Morgan, 1969). The farther fields involve a high labour input of journey-to-work and normally receive fewer visits. The preferred crops on the farther fields are those which can most successfully be produced by extensive methods and are often, where supervision is difficult, of low value. Fields near the settlement may be the chief or even the sole recipients of manure, refuse and night soil brought from the village, livestock pens or a pasture area immediately round the village. An innermost manure zone therefore often occurs, associated with high labour inputs, frequent visits to fields and greater care in husbandry. Land-use zonations may thus include an inner zone of permanent cultivation, a middle zone of various forms of rotational bush fallow and an outer zone of shifting cultivation (Prothero, 1957; Manshard, 1961; Sautter, 1962).

In tropical Africa land-use zonation of this kind has been observed most frequently in the savanna, probably because it has been more difficult to observe in rainforest because of the increased problems of mapping, especially when based on aerial photographs. Also the differences in weeding and clearance, and lack of livestock and consequent lack of manure in rainforest conditions hinder such forms of land-use zonation. Some zonation has, however, been noted (Morgan, 1969B; Grove, 1951). A detailed examination of the spatial organization of agriculture in some north Indian villages has been undertaken by Blaikie (1971), who in addition to noting extensive agriculture at the greater distances from the village and a tendency for intensity of production of any one crop to decrease with increasing distance from any one farm, argued for a principle of movement minimization and also for a wide scatter of fields in each holding in order to ensure that rainfall would be adequate on at least one field. Jackson (1972) has concluded that maximization of returns to labour is the key issue, not returns to land, and even in areas of land shortage, where land may become more important in decision making, rings of cultivation will not necessarily be formed. It is evident in most cases, however, that the rings are formed because labour saving by movement minimization is sought. Land shortage may in fact lead to careful adjustment to soil and drainage differences, which will obscure the ring distribution. For example, land-use maps of Gambia (*Gambia, Land Use Survey*, 1958) show ring patterns more clearly where villages are farther apart and there is land to spare. Whilst intensive agriculture is often induced by population pressure and may even mean poorer returns per unit of input as extra labour is applied, it may provide better returns per unit of labour expended where the alternative is many hours of travel to distant fields. Such intensification is not so much a new development or innovation as an adjustment to maintain a system of production as nearly traditional as possible as the area under cultivation from a given centre expands. Innovation in the sense of a new crop or technique offering increased production and real incomes does not necessarily arrive at or near the centre of such a system, as Boserup (1965) has suggested, although there are many examples of new crops and modern commercial cultivation in the innermost rings, as in southwestern Nigeria, where Oyeleye (1973) contrasted the concentration of flue-cured tobacco around barnsites with the wide scatter of air-cured tobacco. An examination of the land-use maps of Gambia, for example, shows that the main commercial crop, groundnuts, occupies innermost ring locations in some villages and outer ring locations in others. The logic of ring location can differ not only from village to village, according to the range of alternative crop and livestock enterprises, degree of commercialism and tenurial arrangements (in some villages younger farmers may have to look mainly to outer ring locations for agricultural expansion), but from farmer to farmer according to age, family needs, personal preferences and spatial differences (farmers within a single village may have different proportions of holding at equivalent distances from the farm).

PROBLEMS OF MANAGEMENT AND RISK IN SMALL FARMING SYSTEMS

The risks involved in husbandry practice and decision making with regard to the adoption of innovations are considerable on small farms for they may involve a drop in income below the level needed to avoid starvation. Many millions of small farmers are quite unable to consider attaining some desired level of moderate comfort, but instead are wholly preoccupied with survival. The size of farm leaves no room for experiment, and improvement frequently requires not just incentives which must be very large in order to overcome the aversion to the risks, but guarantees against certain degrees of failure. Where new schemes are to be embarked upon without any major trials on the farms concerned, even if trials have already been conducted on government experimental farms, there has to be a high degree of certainty where there is so much to lose. There is moreover, as Blaikie (1971) has remarked in connection with a study of land-use zonation, a spatial configuration of risk. Zonation and fragmentation can be regarded as risk reducing. Chapman (1974) noted the tendency of agricultural research to concentrate either on the national aggregate level or on the level of crop technology and hardly ever on the intervening level of farm management, where the innovation and location decisions must be made. Irrigation in Bihar was seen by Chapman as a risk reducing technique, not simply in the sense that it reduced dependence on uncertain weather, but that it served as a regulator enabling the farmer to make not just one decision, i.e. to plant a crop, but a series of constructive decisions during the period of crop growth. Although irrigation tends to reduce variability of yield and income, there are still risks in water supply, especially in the rationing systems used, involving in one case in northern India a mismatch between water supply and crop needs and considerable difficulties in expanding hybrid grain production, especially on small farms less favoured with adequate water supply (Reidinger, 1974).

Many peasant farmers are seeking ways out of a risk dilemma. One course is to innovate, grow commercial crops and expand productivity either by increasing the holding size or by intensification. Without adequate guarantees, even with inducements this course is open largely to just the more successful farmers whose incomes are well above the starvation line and who fear the risk of failure less. Another course is to retreat into traditionalism in order to ensure some return even if somewhat unsatisfactory. For many small farmers in overcrowded areas there is no other course, so between the two broad groups of farmers there is a tendency to pull apart both economically and socially. A third course is to seek other employment, sometimes part-time or seasonal, in order to continue farming, or to invest in the education and job seeking activities of one or a few members of the family. Where the risks in the shanty towns, great as they are, are hardly worse than the risks on the farm and the prizes are so considerable, can urban migration on the current vast scale be surprising? Nor can it be surprising when prices drop and markets become uncertain that small farmers should prefer to produce food crops for home markets rather than export, and when developing the riskier sole crop systems on rain dependent land should prefer a crop such as cassava which gives a satisfactory yield even on the poorest soils and being 'self-storing', provides a food resource when other staple foodstuffs are in short supply.

In order to build models of the decision making environment of farmers incorporating risk, games theory has been used by both agricultural geographers and economists. The theory is optimizing, that is it assumes that choices should be made so as to maximize expected utility (rather than expected profit) (Edwards, 1954), a somewhat doubtful assumption in much of the context of farming or indeed of most business situations. It would seem likely that farmers seek satisfactory goals rather than maximization. The basic assumptions of von Neuman and Morgenstern in their study of the theory of games and economic behaviour (1944) are:

1. that risky propositions can be ordered in desirability;
2. that the concept of expected utility is behaviourly meaningful;
3. that choices among risky alternatives are made in such a way that they maximize expected utility.

Games theory has been applied to Third World farming situations, more especially by Gould (1963), who applied it to two hypothetical problems in a West African situation, one a market selection and sales allocation problem for cattle traders in Upper Volta, Mali and Niger, and the other a land use allocation problem in a Ghanaian village, in which it was assumed that crop choice was mainly determined by a games strategy concerned only with whether years would be wet or dry. The simplifying assumption was useful to demonstrate method, but the value of the technique in the solution of real problems may be questioned, as Pred (1967) has done. Farmers may not choose the minimax principle with its ultra-defensive tactics, but may take risks in order to increase incomes. Indeed Dillon and Anderson (1971) have even postulated risk preference amongst some farmers in Third World locations. Decisions of this kind depend much on resources. In some areas climatic conditions or rather general weather trends may be fairly predictable from past experience, but in most of the Third World this is not so. Farmers may be more aware of disastrous years than the overall pattern. They may form a simple defensive strategy against rather bad years, being perhaps unable to cope with the worst that can happen, rather than choose crops on the basis of remembered total rainfall occurrences over several years. Weather conditions cannot in fact be divided into simple types, and considerable variations are possible between the total 'patterns' of each year, making the computational problems of a games strategy extremely complex. Again weather or general climatic trends are not the sole factors and in some areas are clearly not the dominant factors in crop choice and land allocation decisions. The interrelationship of the crops themselves in farming operations is in most cases as important a consideration as any other in land allocation (and labour allocation) decisions.

THE SPATIAL STRUCTURE OF FARMING SYSTEMS: 2. LARGE FARMS

Farm units or estate systems in which numerous small farmers are subject to some degree of central management are not simply large; their size permits economies of scale, especially in marketing and the purchase of inputs, and sometimes allows specialization in the use of labour, machinery, modern fertilizers and chemical techniques. They differ not only from smallholdings but from village groups of smallholders where central direction and planning are lacking. In the capitalist world the plantation and the managed estate or latifundium are the approximate equivalents of the socialist state farm and collective farm. The co-operative in all its many forms provides a collection of hybrids, from groups of farmers combining to effect scale economies in marketing or inputs, but acting otherwise independently, to co-operatives in which all farming activities are conducted on a shared basis. All have the attribute of a degree of central direction imposing a regular pattern on the agricultural landscape and normally resulting in a spatial structure with a managerial core, often with a central processing and packing unit, and usually simpler than the spatial structures so far described. Most such units are concerned with the modern commercial sector of the economy, whether capitalist or socialist. Subsistence is left to small independent farmers. They are frequently innovative, even pioneers in experimenting, developing and introducing innovations and in developing markets, chiefly overseas, for their produce. In some situations, as discussed above (pp. 76–7, 108, 143), however, innovation may be regarded as a high cost yielding little comparable return, and many large estates have become agricultural backwaters whose owners prefer to invest their capital elsewhere.

There are many large commercial farm units which are members of international land holding and farming systems. Often they are found to exist in combination with industrial forms of production in what is a vertically integrated system. Many plantations producing industrial raw materials such as rubber are the result of investment in the tropics by large manufacturing concerns. Fordlandia in Brazil and the Firestone and Goodrich estates in Liberia are good examples. Plantations and estates often represent substantial investments in land, allowing for future expansion within a single unit and avoiding the dangers of accretion by fragmentation. Usually a significant portion of the holding is not in the major export crop produced but in subsidiary food crops for the workers, in timber, in pasture and in little used forest or woodland. Large plantation and estate systems belong to an era of abundant land and low labour costs and there are fewer and fewer countries to which, in consequence, they are suited in their traditional form. Nevertheless, large farms, as such, may still be regarded as a most profitable kind of agricultural operation, even in Third World countries whose agricultural economies are largely dominated by the peasantry. Generally in the Third World, despite Government promotions of state and collective farms, increased co-operation between small farmers, and the encouragement of investment from external sources in certain large-scale schemes, the overall impression is of the break-up of large estates and smaller average sizes of farm. This decline is affecting productivity not only in reducing the output of a hitherto important group of producers, but in the tendency of several of those who survive to minimize inputs and avoid replacing old stock in the face of an uncertain political and economic future.

a. *Plantations and estates*. Most plantation and estate layouts are relatively simple, but do not all conform to the notion of a centralized system in which the main crop is close to the centre with its administration and processing, whilst subsidiary activities are located peripherally. In sugar planting, the crop, whether planted by the land owners, by tenants or by independent farmers, must be located close to the mill because with each day's delay the juice in the cane will continue fermenting, with rising losses of sugar yields. In coffee planting in São Paulo the *fazenda* headquarters are normally located together with the housing for the workers and their families, drying platforms and buildings in some central position and in a valley surrounded by food crop land, pasture and unused land. It is coffee that occupies the peripheral position on the ridge tops and valley sides, often on heavier soils than on the valley slopes and on well-drained sites which are free from frost. Estate boundaries are mostly along the ridge tops. Increasing interest in growing more than one commercial crop and making full use of varied terrain has developed as the frontiers of land settlement have been reached and increasing intensification and care in production faces the problems of varying prices on world markets. Diversification hardly matches that experienced on many small farms since the advantages of specialization are still paramount in plantation agriculture. Nevertheless increased interest in experimenting with new crops such as soya beans or supplying national or local markets with fruit, cotton or vegetables has produced a more varied spatial pattern in many plantation systems.

b. *Farm settlement schemes and co-operatives.* One of the key problems in developing agriculture has been seen to be the technical competence of the 'new' farmers and, with that, lack of capital and problems of marketing. A wholesale rather than a piecemeal approach to developing agricultural productivity has been the creation of farm settlements based on various models. One of the more widespread models is that of the Israeli *moshavim* adopted as the basis of new farming schemes in several countries. In Western and Eastern Nigeria, for example, *moshavim*-type farm settlements (see pp. 158–9) were planned in the early 1960s as nearly self-contained units, organized co-operatively, with a degree of central planning, direction and handling of inputs and marketing, but with individually operated holdings. Clearance and tillage were done mechanically on land laid out in strips around central villages which contained all the common services and a central co-operative livestock house (excluding chickens, which were to be kept on the farms in larger units of 96 chickens for each holding). Thorough clearance of forest and woodland together with the geometric layout produced a stark contrast with the existing agricultural landscape. Regularity of land-use pattern on each holding together with the centralization of residence and services reflected the authoritarian nature of the system despite the original intention to form a co-operative settlement. Moreover, another spatial factor, namely the recruitment of settlers from villages no further than 30 miles away, whilst it had obvious social advantages, nevertheless made it easy for dissatisfied settlers to return home. The proximity of nearby villages and the new settlements might have helped to promote a spread of new ideas in farming. In fact there is very little evidence of such a spread and even within 15 miles of two settlements nearly 30 per cent of farmers interviewed said they did not know of the existence of the settlements and very few visited them to view or to seek advice (Roider, 1970).

Many schemes do not employ so large an element of central direction, but usually provide clearance and tillage services setting their own characteristic pattern on the agricultural landscape in the geometry of the layout and often in contour ridging. Most have recommended or prescribed crop rotations and treatments, and pasture and forage arrangements for livestock. Most have central processing and storage and handling of harvested crops. It is useful to compare examples in the survey *Change in Agriculture* edited by Bunting (1970), where studies in India, Guatemala, Zambia, Tanzania, Nigeria, Kenya, Uganda and Bolivia are cited. In Kenya the original Swynnerton Plan of land consolidation, accepted by the Government in 1954, became in effect a vast resettlement scheme with centrally-planned agricultural practice on 'economic' holdings (Lawrance, 1970; Clayton, 1964) consisting originally of some 52 000 square miles but since extended further. In Sri Lanka the reform of the traditional *ande* system of tenant farming, whereby landlords might secure as much as half the crop, by the Paddy Land Act of 1958, not only gave large numbers of peasant farmers title to land, but created a new spatial system of management and advice. Cultivation committees were created in each village to manage cultivation and ensure the supply of certain inputs, more especially fertilizers and chemicals. Decision making was in large part removed from the landlord and tenant to a central village organization. Each committee in effect controlled or managed about 160 hectares of paddy land, aided by agricultural instructors, village cultivation officers and district revenue officers (Karunatilake, 1971, pp. 105–10). In Tanzania the *ujamaa* system of co-operative villages has had a marked spatial effect on cultivation in areas such as Dodoma, for example, where people were urged to leave their scattered homesteads and come together in nucleated villages, resulting in a migration of farmers in 1971. However, practice has been less effective than planned, and by 1973 few villages had reached the stage where even half their land was communally worked (Connell, 1974). Success in developing co-operative farming in the Third World has been chiefly limited to the formation of societies for purchasing inputs or controlling marketing (see above, pp. 140–1). Co-operative farming as such has been slow to develop even in areas with large numbers of very small farms, partly because of doubt about the efficiency of large co-operative units and lack of trust, and partly because of a frequent preference for independence in farming as a way of life, even on a low income. It is, however, probably too early to conclude with Heseltine that co-operatives have generally failed (Heseltine, 1973). For many governments co-operative organization in at least some of the functions of farm management is essential to achieve economies of scale whilst protecting the interests of small farmers. Co-operatives have also been an important means for providing rural credit. Thus the former East Pakistan's Comilla Project, intended as a rural development laboratory to test methods of modernizing agriculture, was organized in village-based co-operatives integrated into one co-operative system (Luykx, 1970).

c. *An irrigated farming scheme—the Sudan Gezira.* Planned in 1904 the Gezira Scheme is one of the oldest 'modern' agricultural developments involving production for an export market, small tenant holdings combining livestock with subsistence and commercial crops, planned production systems and central control of marketing and of many inputs, and a considerable capital investment. The original intention was to provide Sudan with a major export crop in order to raise government revenue. Work was begun on a dam at Sennar on the Blue Nile in 1913 and after experiment on schemes totalling some 20 000 hectares, approximately 95 000 hectares were irrigated when water was supplied by the Main Canal in 1925, increased to 120 000 in the following year. The Gezira, or 'island', is the land between the Blue and the White Niles, mainly the plain of some two million hectares north of the Sennar–Kosti railway, including the Manaqil Ridge in the south centre. The original area of 1925 lies on the eastern side (Fig. 26) close to the Blue Nile, but subsequent extensions to the north and to the southwest in the great arc of the Manaqil Extension increased the area to some 400 000 hectares of irrigated land, and more recently to 755 000 hectares. Water is supplied by gravity flow through a system of canals and distributary feeders designed to allow for rectangular plots of 4 hectares between the water-courses and provided with escape channels and drains to remove excess water. Plots were originally grouped in units of 9, forming a water-control unit, and every third plot provided the cotton portion of a holding for a single tenant, whose irrigation and cultivation practices were subject to central supervision. The low salts content of the Blue Nile meant very little danger of harmful accumulation in the impermeable soils, although water has to be drained, from time to time, from the older eastern scheme and has to be pumped from depressions in the western areas. The water is also relatively free of bacteria, but bilharzia infection is a danger and has to be controlled with copper sulphate (Tothill, 1948; Barbour, 1961).

The general spatial structure tends to be ordered by slope and by considerations of water supply which are partially limited by agreement with Egypt. Locally the layout is determined by the geometry of the canal network and by the contrary pulls of a preference by the settlers for village life and the need for tenants to be near their fields in order to exercise regular supervision over their holdings. The original holdings of 1925 totalled 3 plots of some 12 hectares comprising over 4 in cotton, 2 in *dhura* sorghum, 2 in legumes (*lubia* or *lablab niger*) and 4 in fallow, later increased to just under 17 hectares divided into 4 plots of over 4 hectares each, with one in cotton but two, i.e.

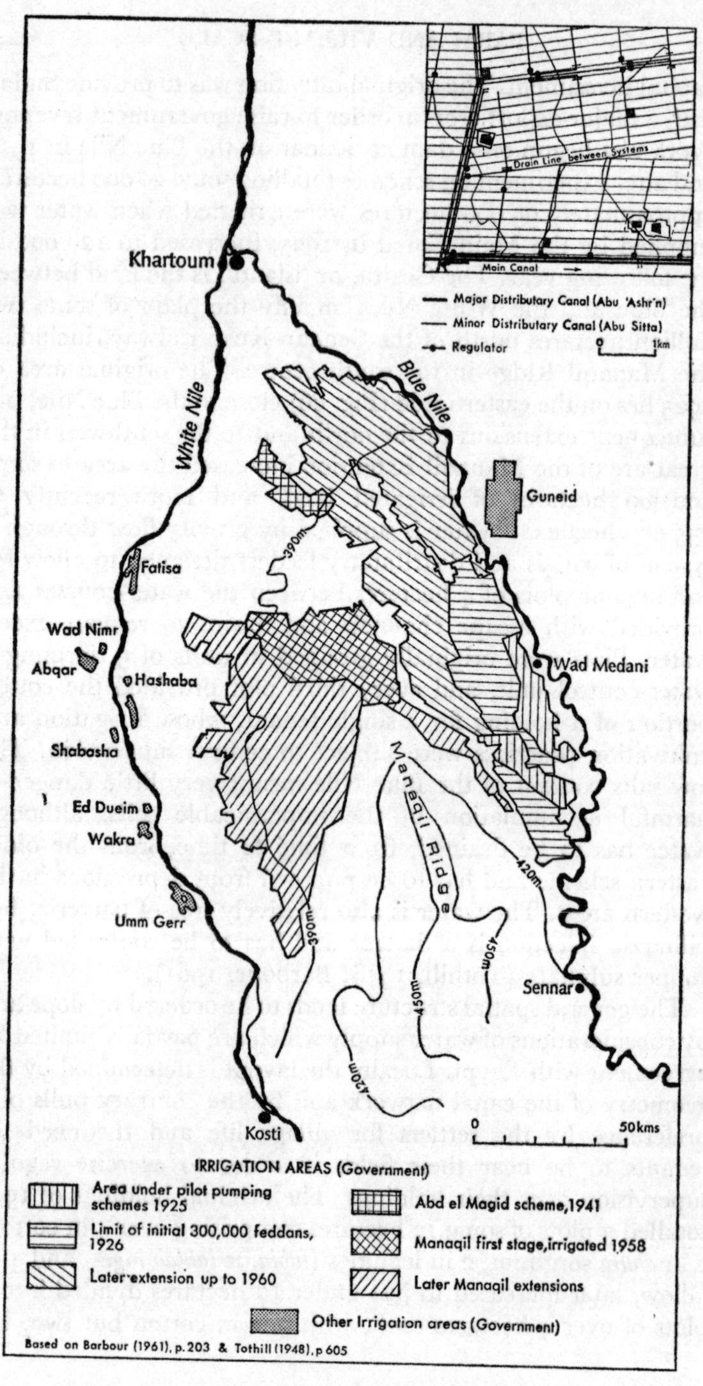

Fig. 26. The Gezira Scheme

over 8 hectares, in fallow, and one in *dhura* or *lubia* or both, or also in fallow. This apparently lavish use of land, was developed in order to maximize cotton returns from a supply of water limited by the Nile Waters Agreement with Egypt, although the size of the irrigation channels in any case limited water flow to only half the available irrigated area at any one time, a situation partially eased in 1972 by modification of the sluices on the Sennar Dam. Much more intensive methods could be used were the farmers able to use all the water available. Pressure to use more water and intensify will increase as holdings become smaller through increased demand by members of the family when plots are divided as tenants die. Intensification will also be encouraged by greater difficulty in hiring labour at busy times of the year—partially eased in recent years by the private purchase of tractors. Many tenancies have been divided into two, and new tenancies in the Manaqil Extension have been reduced to only just over 6 hectares, including some 2 hectares of cotton in a three-course rotation. Fallows have been reduced throughout the area and the average proportion of tenancies in crops has been increased from 44 per cent in 1950 to approximately two-thirds (Thornton, 1972).

Holdings are not compact. The main fields are somewhat separated from each other and the original ideal of close supervision is frequently hard to achieve. The original partnership arrangement between private companies, namely the Sudan Plantations Syndicate Ltd. and the Kassala Cotton Company Ltd., government and tenants has been replaced by a partnership between government, the nationalized Gezira Board and tenants. The government and the Board take 60 per cent of the proceeds from cotton sales or as much as many of the more exacting landlords in tenancy systems cited elsewhere, and much more than in equivalent government controlled schemes in Pakistan and India. The tenant receives 40 per cent together with his food crops and livestock fodder, but compared with other Sudanese farmers has on the whole achieved a fair measure of prosperity, particularly in the cotton boom period of the early 1950s, when the high quality Sudan cottons achieved good prices on the world market. Whilst the tenants bear few risks and the responsibilities of management have been taken from them, the controls over production and the size of holdings together with increased pressure on land and water resources have meant a ceiling to income levels which is low for the ambitious. There has been a general decline in incomes from reduced holdings, with some variation as world market prices have fluctuated. Although modern inputs are used, the farmers themselves have been so guided that in a sense they have not been able to share in the development process. As Barbour (1961, p. 207) has concluded when referring to three towns of the eastern Gezira, 'They are rather smaller than might be expected in a prosperous farming area, the main reason being the unified system of administration and the rigid cropping policy that are imposed by the Gezira Board. This means that individual farmers have neither large agricultural surpluses to dispose of at local markets, nor occasion to borrow money from local banks or to buy seed, machinery, and fertilizers from local merchants.' Apart from tractor purchase this has generally remained true. Thus rigid control, by eliminating many of the main elements in the agricultural infrastructure, has set its own distinctive mark on the spatial structure of the Gezira Scheme.

SPATIAL DISCONTINUITY IN AGRICULTURE: THE PROBLEM OF FRAGMENTATION

Parcellization or the fragmentation of a holding into scattered parcels of land has been regarded for the most part as a major spatial impediment to production and to agricultural improvements, especially to the economic use of labour and the introduction of such modern inputs as tractors. Several attempts have been made to provide measures of parcellization, normally using the parameters of distance apart and size of parcels, but none has been particularly satisfactory owing to the variable relationship of parcellization to the farming systems concerned and the variety of inputs hauled or carried between farmstead and parcels. Problems include risks of damage or deterioration both to inputs and to crops in transit between parcels, difficulties of access, more especially where machinery is involved, design of labour use in a farm system where the maximum advantage has to be made of each visit to a parcel of distinctive size, and problems of control over stealing, of damage by pests and of water supply control where irrigation or floodland are involved.

Fragmentation is a major feature of overcrowded lands, particularly where land is divided by inheritance, conveyed to new owners on marriage or simply bought and sold. It is also a feature of usufructuary systems where some equalization of difficulties such as access and variable soil and slope character is desired, and of successful development where richer farmers wish to increase their holdings size by the purchase of additional properties. In a sense parcellization in the Third World is both good and evil, showing growth as well as decline. It can be removed by reorganization schemes such as the Kenya programme of land reform discussed above (pp. 145–6), but such schemes, whilst they may make possible a superior spatial organization for the use of modern inputs, will rarely eliminate differences between farmers. On the contrary they may well increase them, enabling some farmers to become much more prosperous than others and be encouraged to acquire more land. Nor do they necessarily remove the factors which may once more bring subdivision of holdings and further fragmentation. Alternatives are tight control, usually by government, legislation in favour of primogeniture, which will almost certainly be opposed or nullified by alternative arrangements, or the creation of new forms of tenure, including co-operative farms, in which the farm families hold a share in the business and cannot sell or acquire or bequeath land.

The idea that only compact units permit 'rational' farming or are to be preferred for their net economic gains is generally unsound. Where cultivators from large villages manage smallholdings there are real gains in efficiency of total operation and in minimization of risk to be made from sharing different distances to fields and differences in environmental quality. Land-use zonation (pp. 236–9) and parcellization are related phenomena, and it would be difficult to imagine the elimination of parcellization without the elimination of land-use zonation by population dispersal, which would produce social and other economic losses, notably in marketing arrangements and in the supply of inputs, information and advice. For large villages the elimination of parcellization would usually make no difference to the total amount of walking to fields. Only where the fragmentation process has severely reduced the size of parcels and increased their number and scatter to uneconomic levels are there gains to be made. Thus in certain areas farmed by the Kikuyu in Kenya, for example, fragmentation had reached alarming proportions and consolidation was seen as essential whatever the inequalities and changes involved. Taylor (1969) has cited cases of twenty or more fragments in single holdings in the Fort Hall District. In several Asian countries many more spectacular examples could be cited. Nevertheless it can be argued, as Farmer (1960) has done, that fragmentation and subdivision are symptoms, not a disease in themselves, and that attempts to remove fragmentation and create 'economic'-sized holdings in the name of agricultural improvement or development can result not just in underemployment but in unemployment. Farmer cited the effect of the Land Husbandry Act in Rhodesia in dispossessing thousands of Africans of their communal rights and driving '30 000 men a year' into the towns. In Kenya consolidation and improved farming schemes also have a mounting problem of landlessness and unemployment as their aftermath. Farmer concluded that the main problem, of which subdivision was a symptom, was the particular economic and demographic situation and that if inheritance laws were not the cause of subdivision then nothing was to be gained by tampering with them. More especially in paddy-growing countries 'with their large and underemployed labour forces' there was no inefficiency in subdivision and the only satisfactory cure for the condition lay in the provision of alternative livelihood.

market towns and offer him an incentive to increase agricultural production' (Haswell, 1973, p. xx). But Haswell insists that many towns are just trading centres generating debt and exhibiting 'mushroom growth' or modernization without development (Haswell, 1973, pp. 47–8). Whilst producing some agricultural change on their rural peripheries they tend to exhibit a condition of 'ruralization' themselves as vast numbers move into them in search of a better life. The farmers are caught up in burdens of debt and taxation whilst their sole important resource other than their land, namely labour, is drained away, including in that drain many of the youngest and most enterprising. The effects must not, however, be exaggerated. Not all cities are 'parasitic' nor are so-called parasitic cities entirely engaged in draining rural resources. They frequently provide essential services including transport, marketing and the supply of tools, seeds, fertilizers, machines and information. The incentive of consumer goods to encourage mainly subsistence farmers to increase production, to adopt new crops and even new methods may be essential in the early stage of agricultural transformation. The fact that increase in productivity may need expenditure on capital goods and that substitutes for both land and labour may have to be sought may be the first realization of the need to effect savings and investment in order to raise incomes. Development must obviously be paid for. Whatever its long-term effects, its immediate result may be income reduction, with less money available to farmers as consumers. Their capacity to spend may rise but by less than their effort may have expected in order that savings may be made, or it may even suffer a decline. The development debate must concentrate on how far the evidence for the diversion of rural income and resources into non-profitable activities, or activities which are in no sense an investment, indicates a genuine and extensive parasitism or simply a tendency for sectoral differentials in rates of growth to be temporarily out of gear. Certainly there are instances of spectacular mushrooming of urban development, of top heavy commercial sectors and bureaucracies. Some Latin American countries with traditions of absentee landlordism and tenant farmer mobility have encouraged these to an extent which governments find difficult or impossible to check. There are, however, many cases where rural population increase is still faster than urban and where the incentives of a market economy have hardly begun to penetrate.

The problem of extremes in the organization of space in developing countries, of policies favouring large cities, which as 'incubators of change' are both creative and parasitic, versus policies favouring small villages, which are thought to preserve special social and spiritual values, has been thoroughly discussed by Johnson (1970). He has argued in favour of constructing a 'rational' settlement and services hierarchy in order to improve the connections between rural areas and cities, to encourage the creation of 'truly functional economic areas' and adequate competition in the markets amongst buyers, and also to provide a settlement pattern which will favour democratic processes. The idea that there are gaps in many settlement hierarchies (more especially that there are too few 'medium-sized' towns), which limit or even prevent contact between rural areas and large cities is important. In many cases, however, where contact has been developed, such contact has served only to encourage a flow of goods and people away from the rural areas, hindering agricultural progress and exacerbating the problems of urban poverty. Thus the planning of settlements, however well ordered, will not in itself solve these problems, particularly where the current economic and social incentives favour city living so heavily and cause farmers to despise their occupational status and find only poor rewards for considerable effort. The creation of adequate rural transport and market systems such as the regulated markets of India (Johnson, 1970, pp. 101–11) with proper attention to rewards for effort and to the problems of rural capital formation are essential prerequisites in any plan which seeks to slow down the pace of urbanward movement and encourage higher agricultural productivity.

Rural labour 'surplus', as understood with all its structural problems, or rather low-productivity labour associated with seasonal surplus, is still a major difficulty, accompanied by low incomes and hindering the introduction of labour saving techniques and consequent higher output per man. The difficulty is likely to increase in many countries where the rates of growth in industrial production required to cope with the situation are at present quite beyond the capacity of conceivable investment and availability of resources. Although 'surplus' occurs yet farming may suffer from seasonal labour shortages and disruption of agricultural production wherever agriculture loses labour to some alternative industry. As concluded earlier (p. 164), mechanization in a labour surplus environment is something of a paradox. There is the possibility that mechanization may result not so much from the desire to raise overall output per man as from the need to replace a local labour force by machines in order to ensure satisfactory control in a new crop demand situation. Thus Jacoby has seen the increasing use of machinery in the Punjab, Bihar and central Luzon as evidence of economic and social disintegration, reflecting in part rising land prices, decreasing profitability of farm tenancies and the growing preference of landowners to take over tenant farmers and incorporate them into vast holdings operated by machines, supplemented by seasonal labour. He has seen this in several instances as encouraged by the operations of transnational corporations which offer contracts to the larger landowners or to government for the large-scale production of industrial or luxury crops for export (Jacoby, 1975).

As Day and Singh showed in the Punjab by means of a model of the agricultural economy, peasant farmers are amenable to economic incentives and respond rationally to them, even though as decentralized decision makers they suffer from ignorance and difficulties in achieving an adequate understanding of the factors involved (Day and Singh, 1975). Agriculture in consequence can develop rapidly within the framework of a market oriented economy, providing that appropriate developments are facilitated outside the farm sector and more especially that such developments are appropriately 'balanced' or are in a suitable equilibrium to avoid bottlenecks in the agricultural system. But all these developments in infrastructure provide only the necessary conditions for agricultural development. They are essential, but as a framework, and provide no guarantee in themselves that development will take place. The sufficient conditions for agricultural progress are still direct demand incentives, enterprise, the desire to raise output and the availability of resources to do so. Moreover, both necessary and sufficient conditions must be appraised in their spatial context. Not only must development cope with environmental variation and the location-specific nature of many of the techniques required, but even if all other conditions were uniform the problems of distance and accessibility would create differences in development between one place and another. In the 'uniform plain' of theoretical geography development must therefore create its own spatial pattern. Spatial 'inequality' is a part of the diffusion process and a product of the varying nature both of demand and of the infrastructures which must serve agriculture. Not all farmers can be large-scale grain farmers. Road networks, however well designed, inevitably impose different costs of accessibility to markets and resources on different farmers. The differences can be minimized and systems can be designed to ensure special encouragement of the poorer farmers, depending on the extent to which investment resources are available. Regional planning is an essential part of the process wherever governments are concerned to encourage agricultural transformation, but so far there are few cases where regional planning has done more than examine environment and attempt to define regional potential. To the map of potential regions we need to fit a map of regional development structures and incorporate the lessons of space that Thünen tried in effect to teach one hundred and fifty years ago.

As this text has tried to show, spatial inequality is a feature which needs appreciation at a variety of scales. These include world aspects where planning is impossible and where results can be achieved only by international agreement. The pace of development and its conditions vary enormously within the Third World and agricultural progress cannot be seen in isolation from changes in internal trade and in political relationships. The difficulties of world inflation and more especially of the rising costs of inputs such as oil and fertilizers, which in several countries are vital resources for agricultural transformation, threaten further advance for many of the economies concerned or may even bring falls in productivity. The rush to develop has in some ways been its own undoing as the bids for scarce resources have risen faster than the capacity to make them available. In some cases, as for example, in the fertilizer industry, they have been attended by other problems such as the threat of pollution or failure in application through ecological disaster, most commonly in the early 1970s through drought. With rising trade gaps and rising costs have come increasing burdens of international debt for many Third World countries, resulting in dependence on other countries for the resources needed even for survival. It is often hard to counter Myrdal's argument that 'scientific and technological advance in the developed countries has had, and is now having, an impact on the underdeveloped countries which, on balance, is detrimental to their development prospects' (Myrdal, 1971, p. 55). Haswell's ideas of the institutionalization of poverty in the Third World go further. She argues that development planning accompanies an excessive preoccupation with world food shortages. Short order massive development programmes intended to protect against natural calamities may result instead in a total breakdown of survival mechanisms. This suggests that even attempts at aid may be detrimental because of failure to understand the motives for production of Third World farmers (Haswell, 1975, pp. 207–8). International relationships are for the present such that development in some countries may be seen as causing 'underdevelopment' or rather poverty in others. We may accept Myrdal's conclusion that underdevelopment is not just a lack of change, but a product of the relationships involved, and that development will not inevitably occur by stages as investment proceeds. 'The usual view that differences in levels of development have only a "dimensional", not a "qualitative" character, and more specifically that there is only a "time lag" between developed and underdeveloped countries —which, like much else in the post-war approach, goes back to Marx—is mistaken. As these thoughts have been developed in the so-called theory of the "stages of growth", they are based on metaphysical preconceptions of the teleological variety' (Myrdal, 1971, p. 58). This does not mean, however, that hope of development has to be rejected, only that a particular theory of development is no longer tenable and that there is no inevitability about the development process and the elimination of poverty even with suitable injections of capital. Hence the importance of a scale view of development in its geographical relationships in order to understand and order the very different situations which have been created and for which we may attempt to evolve very different yet related formulae.

Bibliography

ABERCROMBIE, K. C. (1975): 'Agricultural mechanization and employment in developing countries', *FAO Monthly Bull. of Agric. Econ. and Stat.*, 24 (5), 1–9.

AJAEGBU, H. I. (1970): 'Food crop farming in the coastal area of southwestern Nigeria', *Journ. Trop. Geogr.*, 31, 1–9.

AKINTOYE, S. A. (1969): 'The Ondo road eastwards of Lagos c. 1870–95', *Journ. Afr. Hist.*, 10 (4), 581–98.

ALLAN, J. A. and ROSING, K. E. (1973): *Disparities in the recognition of indicators of agricultural development—a North West Indian case*, Paper presented to the Inst. Brit. Geogr. Developing Areas Study Group Symposium, September 1973.

ALLAN, M. (1967): *The Hookers of Kew*, London.

ALLAN, W. (1965): *The African husbandman*, Edinburgh.

ALTSCHUL, A. M. and ROSENFIELD, D. (1970): 'Protein supplementation: satisfying man's food needs', *Progress*, 54 (3), 76–84.

ANTONINI, G. A. (1971): 'Peasant agriculture in Northwestern Dominican Republic', *Journ. Trop. Geogr.*, 32, 1–10.

AXINN, G. H. (1972): *Modernizing world agriculture: a comparative study of agricultural extension education systems*, New York.

BAIROCH, P. (1975): *The economic development of the Third World since 1900* (transl. by C. Postan), London.

BAKER, O. E. (1926–32): 'Agricultural regions of North America', *Econ. Geogr.*, vols. 2–8.

BALANDIER, G. (1972): *Political anthropology*, Harmondsworth, transl. by A. M. Sheridan Smith from 'Anthropologie politique' (Paris, 1967).

BALDWIN, K. D. S. (1957): *The Niger Agricultural Project*, Oxford.

BALDWIN, R. E. (1956): 'Patterns of development in newly settled regions', *Manch. Schl. of Econ. and Soc. Studies*, 24, 161–79.

BALDWIN, R. E. (1966): *Economic development and export growth: a study of Northern Rhodesia, 1920–1960*, Berkeley and Los Angeles.
BARBOUR, K. M. (1961): *The Republic of the Sudan*, London.
BARDHAN, P. K. (1970): 'On the minimum level of living and the rural poor', *Indian Econ. Rev.*, 5(1), 129–36.
BARDHAN, P. K. (1973): 'On the incidence of poverty in rural India', *Econ. and Polit. Weekly*, Bombay, 8, 245–54.
BAREN, F. A. VAN (1960): 'Soils in relation to population in tropical regions', *Tijds. voor Econ. en Soc. Geogr.*, 51 (9), 230–4.
BARKER, R. (1970): 'The contribution of the International Rice Research Institute to Asian agricultural development', in A. H. Bunting (ed.), *Change in agriculture*, London, 207–18.
BARRAZA ALLANDE, L. (1974): 'Regional agricultural growth and economic development', in N. Islam (ed.), *Agricultural policy in developing countries*, London, 363–87.
BARTLETT, H. H. (1955, 1957, 1961): *Fire in relation to primitive agriculture and grazing in the tropics*, I–III, Ann Arbor.
BATH, B. H. SLICHER VAN (1963): *The agrarian history of Western Europe A.D. 500–1850*, transl. by Olive Ordish, London.
BAUER, P. T. (1954): *West African trade*, London.
BECKFORD, G. L. (1969): 'The economies of agricultural resource use and development in plantation economies', *Soc. and Econ. Studies, Jamaica*, 18, 321–47.
BEGUIN, H. (1964): *Modèles géographiques pour l'espace rural african*, Acad. Roy. des Sciences d'Outre-Mer, Brussels.
BELL, P. F. (1969): 'Thailand's North East regional under development, insurgency and official response', *Pacif. Aff.*, 42, 47–54.
BELSHAW, H. (1959): *Agricultural credit in economically underdeveloped countries*, Rome, FAO Agricultural Studies, No. 46.
BERRY, S. S. (1968): 'Christianity and the rise of cocoa-growing in Ibadan and Ondo', *Journ. Hist. Soc. Nigeria*, 4 (3), 439–51.
BERRY, S. S. (1974): 'The concept of innovation and the history of cocoa farming in Western Nigeria', *Journ. Afr. Hist.*, 15 (1), 83–95.
BIRD, R. M. (1974): *Taxing agricultural land in developing countries*, Cambridge, Mass.
BLAIKIE, P. M. (1971): 'Spatial organization of agriculture in some north Indian villages', *Trans. Inst. Brit. Geogr.*, Pt. I, 52, 1–40; Pt. II, 53, 15–30.
BOEKE, J. H. (1953): *Economics and economic policy of dual societies as exemplified by Indonesia*, New York.
BOINVILLE, C. A. C. DE (1968): 'World malnutrition and the role of animal feeds', *Progress*, 52 (2), 172–8.

BOSERUP, E. (1965): *The conditions of agricultural growth: the economics of agrarian change under population pressure*, London.

BOSERUP, E. (1974): 'Food supply and population in developing countries: present status and prospects', in N. Islam (ed.), *Agricultural policy in developing countries*, London, Conference of the International Economic Association in 1972, 164–76.

BROMLEY, R. J. (1971): 'Markets in the developing countries: a review', *Geogr.*, 56 (2), 124–32.

BROOKFIELD, H. C. (1961): 'The highland peoples of New Guinea: a study of distribution and localization', *Geogr. Journ.*, 127 (4), 436–48.

BROOKFIELD, H. C. (1962): 'Local study and comparative method: an example from central New Guinea', *Ann. Assoc. Amer. Geogr.*, 52 (3), 242–54.

BROOKFIELD, H. C. (1964): 'The ecology of highland settlement: some suggestions', *Amer. Anthrop.*, 66 (4), 20–38 and 309–22.

BROOKFIELD, H. C. (1973): 'Introduction: explaining or understanding', in Brookfield (ed.), *The Pacific in transition: geographical perspectives on adaptation and change*, London, 3–22.

BROOKFIELD, H. C. (1975): *Interdependent development*, London.

BROWN, D. D. (1971): *Agricultural development in India's districts*, Cambridge, Mass.

BROWN, L. R. (1970): *Seeds of change: the green revolution and development in the 1970s*, New York.

BROWN, P. and BROOKFIELD, H. C. (1959): *Chimbu land and society*, Sydney.

BUNTING, A. H. (ed.) (1970): *Change in agriculture*, London.

BUNTING, A. H. (1970): 'Review and conclusions', in A. H. Bunting (ed.), *Change in agriculture*, International seminar on change in agriculture, University of Reading, London, 715–93.

BURLEY, T. M. (1973): 'Brazilian agriculture expanded production in soya beans only in 1972', *World Crops*, 25 (5), 243.

BURLEY, T. M. (1974): 'Mechanization of Asian agriculture: its potential and pitfalls', *World Crops*, 26 (2), 88.

BURROUGH, J. B. (1973): 'Ethnicity as a determinant of peasant farming characteristics: the canals polder, Guyana', *Journ. Trop. Geogr.*, 37, 1–8.

CAMERON, H. C. (1952): *Sir Joseph Banks*, London.

CHAKRAVARTI, A. K. (1971): 'Changes in the patterns of foodgrain production and sufficiency level in India, 1921 to 1951', *Journ. Trop. Geogr.*, 32, 11–30.

CHANCELLOR, W. J. (1970): *Survey of tractor contractor operations in Thailand and Malaysia*, Davis, California.

CHANDRA, S., BOER, A. J. DE and EVENSON, J. P. (1974): 'Economics and energetics: Sigatoka Valley, Fiji', *World Crops*, 26 (1), 34–7.

CHAPMAN, G. P. (1974): 'Perception and regulation: a case study of farmers in Bihar', *Trans. Inst. Brit. Geogr.*, 62, 71–93.

CHENERY, H. B. (1955): 'The role of industrialization in development programmes', *Amer. Econ. Rev.*, 45 (2), 40–57.

CHISHOLM, M. (1962): *Rural settlement and land use*, London.

CHISHOLM, M. (1964): 'Problems in the classification and use of farming-type regions', *Trans. Inst. Brit. Geogr.*, 35, 91–103.

CLARK C. (1940): The morphology of economic growth, in Clark, C. (ed.) *The conditions of economic progress*, London, 337–73.

CLARK, C. (1953): 'Population growth and living standards', *Internat. Lab. Rev.*, 68 (2), 99–117.

CLARK, C. and HASWELL, M. R. (1964): *The economics of subsistence agriculture*, London.

CLARK, R. J. (1968): 'Land reform and peasant market participation on the north Highlands of Bolivia', *Land Econ.*, 44, 153–72.

CLAYTON, E. S. (1964): *Agrarian development in peasant economies: some lessons from Kenya*, Oxford.

CONKLIN, H. C. (1954): 'An ethnoecological approach to shifting agriculture', *Trans. N.Y. Acad. of Sci.*, Series II, 17, 133–42.

CONKLIN, H. C. (1957): 'Population-land balance under systems of tropical forest agriculture', *Symposium of the Ninth Pacific Science Congress*, Bangkok.

CONNELL, J. (1973): 'The geography of development or the development of geography', *Antipode*, 5 (2), 27–39.

CONNELL, J. (1974): 'The evolution of Tanzanian rural development', *Journ. Trop. Geogr.*, 38, 7–18.

COURSEY, D. G. and HAYNES, P. H. (1970): 'Root crops and their potential as food in the tropics', *World Crops*, 22 (4), 261–5.

COURTENAY, P. P. (1971): *Plantation agriculture*, London.

CURTIN, P. D. (1965): *The image of Africa*, London.

DATOO, B. A. (1973): *Population density and agricultural systems in the Uluguru Mountains, Morogoro District*, Bureau of Resource Assessment and Land Use Planning, Univ. of Dar-es-Salaam, Research Paper No. 26.

DAVIES, H. R. J. (1964): 'An agricultural revolution in the African tropics: The development of mechanised agriculture on the clay plains of the Republic of Sudan, *Tijds. voor Econ. en Soc. Geogr.*, 55, 101–8.

DAVIS, D. H. (1948): *The earth and man* (revised edn.), New York.

DAY, R. H. and SINGH, I. (1975): 'A dynamic model of regional agricultural development', *Internat. Reg. Sci. Rev.*, 1 (1), 27–48.

Desai, D. K. (1969): 'Intensive agricultural districts programme', *Econ. and polit. weekly, Bombay*, 4 (26), 83–90.

De Wilde, J. C. (1967): *Experiences with agricultural development in Tropical Africa*, 2 vols., Baltimore.

Dillon, J. L. and Anderson, J. R. (1971): 'Allocative efficiency, traditional agriculture and risk', *Amer. Journ. of Agric. Econ.*, 53 (1), 26–32.

Dixon, C. J. (1975): *Rural income disparities and instability in North-East Thailand*, Paper presented to the Inst. Brit. Geogr. Areas Study Group Symposium, January 1975, Oxford.

Dobby, E. H. G. (1955): 'Padi landscapes of Malaya', *Malayan Journ. Trop. Geogr.*, 6.

Due, J. M. and Gehring, D. C. (1973): 'Jamaica's strategy for import substitution of vegetables in the 1960s', *Illinois Agric. Econ.*, 13 (1), 20–6.

Dumont, R. (1957): *Types of rural economy* (transl. Magnin), London.

Dumont, R. (1966): *African agricultural development*, FAO, UNO Econ. Com. for Africa, New York.

Dutt, A. K. (1972): 'Daily influence area of Calcutta', *Journ. Trop. Geogr.*, 35, 32–9.

Edwards, W. (1954): 'The theory of decision making', *Psychol. Bull.*, 51 (4), 380–417.

Eicher, C. K. (1970): *Research on agricultural development in five English-speaking West African countries*, Agricultural Development Council Inc., New York.

Elkan, W. (1973): *An introduction to development economics*, Harmondsworth.

Engmann, E. V. T. (1973): 'Spatial convergence-divergence: a quantitative approach to problems of delineating the region', *Journ. Trop. Geogr.*, 37, 16–29.

Evenson, R. E. and Kislev, Y. (1975): 'Investment in agricultural research and extension: a survey of international data', *Econ. dev. and cult. change*, 23 (3), 507–21.

FAO See United Nations.

Farmer, B. H. (1960): 'On not controlling subdivision in paddy-lands', *Trans. and papers, Inst. Brit. Geogr.*, 28, 225–35.

Faulkner, O. and Mackie, J. (1933): *W. African agric.*, Cambridge.

Fei, J. C. H. and Ranis, G. (1964): *Development of the labor surplus economy*, Homewood, Ill.

Fisher, A. G. B. (1945): *Economic progress and social security*, London.

Fisher, C. A. (1964): *South-east Asia*, London.

Floyd, B. (1959): *Changing patterns of African land use in Southern Rhodesia*, Salisbury.

FOGG, W. (1932): 'The suq: a study in the human geography of Morocco', *Geogr.*, 17, 257–8.

FORMAN, S. and RIEGELHAUPT, J. F. (1970): 'Market place and market system: towards a theory of peasant economic integration', *Compar. Stud. in Soc. and Hist.*, *The Hague*, 12 (2), 188–212.

FORTT, J. M. and HOUGHAM, D. A. (1973): 'Environment, population and economic history', in Richards, A. I., Sturrock, F. and Fortt, J. M. (eds.), *Subsistence to commercial farming in present-day Buganda: an economic and anthropological survey*, Cambridge, 17–46.

FOSBERG, F. R., GARNIER, B. J. and KUCHLER, A. W. (1961): 'Delimitation of the humid tropics', *Geogr. Rev.*, 51 (3), 333–47.

FRANK, A. G. (1966): 'The development of underdevelopment', *Monthly Review*, 18 (4), 17–30.

FRANKEL, F. R. (1971): *India's green revolution*, Princeton, New Jersey.

FRANKLIN, S. H. (1962): 'Reflections on the peasantry', *Pac. Viewp.*, 3, 1–26.

FREEMAN, J. D. (1955): *Iban agriculture*, London.

FREIVALDS, J. (1973): 'Agro-industry in Africa', *World Crops*, 25 (3), 124–6.

FRIEDMANN, J. (1966): *Regional development policy: a case study of Venezuela*, Cambridge, Mass.

FRIEDMANN, J. (1972): 'A general theory of polarized development', in Hansen, N. M. (ed.), *Growth centers in regional economic development*, New York, 82–107.

FRIEDMANN, J. and ALONSO, W. (eds.) (1964): *Regional development and planning: a reader*, Cambridge, Mass.

FUNNELL, D. C. (1973): *The geographical analysis of small urban centres and contributions to development problems*, Paper presented to Inst. Brit. Geogr. Developing Areas Study Group Conference on Problems in Relating Development Research to Development.

FÜRER-HEIMENDORF, C. VON (1952): 'Ethnographic notes on some communities of the Wynad', *The Eastern Anthropologist*, 6, 18–36.

FURNIVALL, J. (1945): 'Some problems of tropical economy', in Hinden, R. (ed.), *Fabian colonial essays*, London.

FURNIVALL, J. (1948): *Colonial policy and practice*, London.

GAMBIA, Land Use Survey (1958), 1:25,000 scale sheets, DOS 3001 (Series G823).

GEDDES, W. R. (1973): 'The opium problem in Northern Thailand', in R. Ho and E. C. Chapman (eds.), *Studies of contemporary Thailand*, ANU Research School of Pacific Studies, Dept. of Human Geogr., Publicn. HG/8, 213–34.

GEERTZ, C. (1968): *Agricultural involution: the processes of ecological change in Indonesia*, Berkeley and Los Angeles.

GEORGE, A. R. (1973): 'Processes of sedentarization of nomads in Egypt, Israel and Syria: a comparison', *Geogr.*, 58 (2), 167–9.

GILBERT, A. (1974): *Latin America: a geographical prospective*, Harmondsworth.

GLEAVE, M. B. (1966): 'Hill settlements and their abandonment in tropical Africa', *Trans. Inst. Brit. Geogr.*, 40, 39–49.

GLEAVE, M. B. and WHITE, H. P. (1969): 'Population density and agricultural systems in West Africa', in Thomas, M. F. and Whittington, G. W. (eds.), *Environment and land use in Africa*, London, 272–300.

GLOVER, J., ROBINSON, P. and HENDERSON, J. P. (1954): 'Provisional maps of the reliability of rainfall in East Africa', *Quart. Journ. Roy. Met. Soc.*, 80, 602–9.

GORDON, J. (1971): 'Mechanisation and the small farmer: the need for a broader approach to the problems in West Africa', *World Crops*, 23 (5), 250–1.

GOULD, P. R. (1963): 'Man against his environment: a game theoretic framework', *Ann. Assoc. Amer. Geogr.*, 53, 290–7.

GOUROU, P. (1961): *The tropical world*, third edition, London, originally transl. by E. D. Laborde from Les pays tropicaux, Paris, 1947.

GRIFFIN, E. (1973): 'Testing the von Thünen theory in Uruguay', *Geogr. Rev.*, 63 (4), 500–16.

GRIGG, D. (1970): *The harsh lands: a study in agricultural development*, London.

GRIGG, D. B. (1974): *The agricultural systems of the world: an evolutionary approach*, Cambridge.

GRIGG, D. B. (1975): 'The world's agricultural labour force', *Geogr.*, 60 (3), 194–202.

GRILICHES, Z. (1960): 'Hybrid corn and the economics of innovation', *Science*, 132 (3422), 275–80.

GROVE, A. T. (1951): 'Soil erosion and population density in south-east Nigeria', *Geogr. Journ.*, 117, 291–306.

GROVE, A. T. (1961): 'Population densities and agriculture in Northern Nigeria', in Barbour, K. M. and Prothero, R. M. (eds.), *Essays on African population*, London, 115–36.

GULLIVER, P. H. (1954): 'Jie agriculture', *Uganda Journ.*, 18, 65–70.

GUPTA, S. (1973): 'The role of the public sector in reducing regional disparity in Indian plans', *Journ. Devt. Stud.*, 9(2), 243–60.

GURNAH, A. M. (1973): 'Large scale maize production in Ghana', *World Crops*, 25 (6), 308–11.

HAFNER, J. A. (1973): 'The spatial dynamics of rice milling and commodity flow in central Thailand', *Journ. Trop. Geogr.*, 37, 30–8.

HAGERSTRAND, T. (1953): *Innovationsförloppet ur korologisk synpunkt*, Lund translated as *Innovation diffusion as a spatial process*, by A. Pred, 1967, Chicago.

HAGGETT, P. (1965A): *Locational analysis in human geography*, London.

HAGGETT, P. (1965B): 'Scale components in geographical problems', in Chorley, R. J. and Haggett, P. (eds.), *Frontiers in geographical teaching*, London, 164–85.

HAGGETT, P., CHORLEY, R. J. and STODDART, D. R. (1965): 'Scale standards in geographical research: a new measure of area magnitude', *Nature*, 205, 844–7.

HAILEY, LORD (1957): *An African Survey: revised to 1956*, London.

HALL, P. (ed.) (1966): *Von Thünen's Isolated State* (an English edition of J. H. von Thünen, Der isolierte Staat, 1826), Oxford.

HANSEN, B. (1969): 'Employment and wages in rural Egypt', *Amer. Econ. Rev.*, 59 (3), 298–313.

HARRISS, B. (1974): 'The role of Punjab wheat markets as growth centres, *Geogr. Journ.*, 140 (1), 52–71.

HART, K. (1973): 'Informal income opportunities and urban employment in Ghana, Conference on urban unemployment in Africa', Inst. of Dev. Stud., Univ. Sussex, 1971, in Jolly, de Kadt, Singer and Wilson (1973), 66–70.

HARVEY, D. W. (1968): 'Pattern, process and the scale problem in geographical research', *Trans. Inst. Brit. Geogr.*, 45, 71–8.

HARVEY, D. W. (1969): *Explanation in geography*, London.

HASWELL, M. R. (1973): *Tropical farming economics*, London.

HASWELL, M. R. (1975): *The nature of poverty: a case-history of the first quarter-century after world war II*, London.

HAYAMI, Y. and RUTTAN, V. W. (1971): *Agricultural development: an international perspective*, Baltimore.

HEADY, E. O. (1952): *Economics of agricultural production and resource use*, Englewood Cliffs, N.J.

HEADY, E. O. and DILLON, J. I. (1961): *Agricultural production functions*, Ames, Iowa.

HEATON, L. E. (1969): *The agricultural development of Venezuela*, New York.

HELLEINER, G. K. (1970): 'The fiscal role of the marketing boards in Nigerian economic development, 1947–61', in M. C. Taylor (ed.), *Taxation for African development*, London, 414–51.

HELLEINER, G. K. (1972): *International trade and economic development*, Harmondsworth.

HESELTINE, N. (1973): 'Socialism and development', review of R. Dumont and M. Mazoyer, Socialism and development (London, 1973), *World Crops*, 25 (6), 319–20.

HIGGINS, B. (1959): *Economic development: principles, problems and policies*, New York.

HIGGINS, B. (1972): 'Regional interaction, the frontier and economic growth', in A. R. Kuklinski (ed.), *Growth poles and growth centres in regional planning*, UN Geneva, 263–302.

HILL, P. (1963): *The migrant cocoa-farmers of southern Ghana: a study in rural capitalism*, Cambridge.

HILL, P. (1972): *Rural Hausa: a village and a setting*, Cambridge.

HILL, R. D. (1969): 'Pepper growing in Johore', *Journ. Trop. Agric.*, 28, 32–9.

HILLING, D. (1969): 'The evolution of the major ports of West Africa', *Geogr. Journ.*, 135 (3), 365–78.

HIRSCHMAN, A. O. (1958): *The strategy of economic development*, New Haven.

HO, R. (1969): 'Rice production in Malaya: a review of problems and prospects', *Journ. Trop. Geogr.*, 29, 21–32.

HODDER, B. W. (1965): 'Some comments on the origins of traditional markets in Africa south of the Sahara', *Trans. and Pap., Inst. Brit. Geogr.*, 36, 97–105.

HOPKINS, A. G. (1973): *An economic history of West Africa*, London.

HOSELITZ, B. (1964): 'Social stratification and economic development', *Internat. Soc. Sci. Journ.* (UNESCO), 16 (2), 237–51.

HUANG, Y. (1975): 'Tenancy patterns, producting, and rentals in Malaysia', *Econ. Dev. and Cult. Change*, 23 (4), 703–18.

HUMPHREY, D. H. (1970): 'Comment on E. R. Watts, Commerce and extension (1969)', *World Crops*, 22 (2), 114–15.

INTERNATIONAL LABOUR OFFICE (1971): *Matching employment opportunities and expectations. A programme of action for Ceylon*, Geneva, vol. 1, 20–33 and 117–20.

ISSAWI, C. (1963): *Egypt in revolution*, London.

IZIKOWITZ, K. G. (1951): 'Lamet: hill peasants in French Indochina', *Etnol. Studier*, 17.

JACK, D. T. (1958): *Economic survey of Sierra Leone*, Freetown.

JACKSON, R. (1972): 'A vicious circle?—The consequences of von Thünen in tropical Africa', *Area*, 4 (4), 258–61.

JACOB, A. and UEXKULL, H. VON (1960): *Fertilizer use* (2nd edn.) (transl. by C. L. Whittles), Hannover.

JACOBY, E. H. (1975): 'Transnational corporations and Third World agriculture', *Econ. Dev. and Cult. Change*, 6 (3), 90–7.

James, P. E. (1932): 'The coffee lands of southeastern Brazil', *Geogr. Rev.*, 22, 225–44.

James, P. E. (1959): *Latin America*, London (3rd edn.); (1942, 1st edn.).

Jayaprakash, R. K. (1973): 'The technological breakthrough in agriculture and its possible socio-economic impact in India', *World Crops*, 25 (2), 78–84.

Johnson, B. L. C. (1972): 'Recent developments in rice breeding and some implications for tropical Asia', *Geogr.*, 57 (4) 307–20.

Johnson, E. A. J. (1965): *Market towns and spatial development in India*, New Delhi.

Johnson, E. A. J. (1970): *The organization of space in developing countries*, Cambridge, Mass.

Johnson, G. L. (1968): 'Factor markets and economic development', in McPherson, W. W., *Economic development of tropical agriculture*, Gainsville, 93–111.

Johnston, B. F. (1958): *The staple food economies of western tropical Africa*, Stanford.

Johnston, B. F. and Mellor, J. W. (1961): 'The role of agriculture in economic development', *Amer. Econ. Rev.*, 51 (4), 566–93.

Johnston, R. J. (1973): *Spatial structures*, London.

Jolly, R., De Kadt, E., Singer, H. and Wilson, F. (eds.) (1973): *Third World employment: problems and strategy*, Harmondsworth.

Jones, G. E. (1965): 'Discussion of the geographical typology of agriculture', *Agricultural Geography*, 59–74, IGU Symposium of 1964, Dept. of Geography, University of Liverpool.

Jordan, H. D. (1954): 'The development of rice research in Sierra Leone', *Trop. Agric.*, 31, 27–32.

Jorgenson, D. W. (1966): 'Testing alternative theories of the development of a dual economy', in Adelman, I. and Thorbecke, E. (eds.), *The theory and design of economic development*, Baltimore.

Jorgenson, D. W. (1969): 'The role of agriculture in economic development: classical versus neoclassical models of growth', in Wharton Jr., C. R. (ed.) (1969), 320–48.

Kaldor, N. (1970): 'Taxation for economic development', in M. C. Taylor, *Taxation for African economic development*, London, 158–77.

Karr, G. L., Njoku, A. O. and Kallon, M. F. (1972): 'Economics of the upland and the inland valley swamp rice production systems in Sierra Leone', *Illinois Agric. Econ.*, 12 (1), 12–17.

Karunatilake, H. N. S. (1971): *Economic development in Ceylon*, New York.

Katzman, M. T. (1975): 'Regional development policy in Brazil:

the role of growth poles and development highways in Goias', *Econ. Dev. and Cult. Change*, 24 (1), 75–107.

KAY, G. (1967), *A social geography of Zambia*, London.

KHAN, A. U. (1972): 'Agricultural mechanization: the tropical farmer's dilemma', *World Crops*, 24 (4), 208–13.

KIRBY, J. M. (1974): 'Venezuela: land reform and agricultural development', *World Crops*, 26 (3), 118–21.

KOLAWOLE, M. I. (1973): 'Why Western Nigerian farmers are against cotton growing', *World Crops*, 25 (4), 179–81.

KUZNETS, S. (1959): *Six lectures on economic growth*, Glencoe, Ill.

LA-ANYANE, S. (1974): 'Some constraints on agricultural development in Ghana', in N. Islam (ed.), *Agricultural policy in developing countries*, London, Conference of the International Economic Association in 1972, 388–410.

LASUEN, J. R. (1969): 'On growth poles', *Urban Studies*, 6, 20–49.

LAWRANCE, J. C. D.: 'Land consolidation and registration in Kenya', in A. H. Bunting (ed.), *Change in agriculture*, International seminar on change in agriculture, University of Reading, London, 451–60.

LEMON, A. (1975): *The Indian communities of East Africa and the West Indies*, IBG symposium, Jan. 1975, Birmingham.

LEWIS, W. A. (1954): 'Economic development with unlimited supplies of labour', *Manch. School of Econ. and Soc. Stud.*, 22, 139–91.

LEWIS, W. A. (1955): *Theory of economic growth*, London.

LINTON, D. L. (1961): *The tropical world: an inaugural lecture*, Birmingham.

LIPTON, M. (1968): 'Strategy for agriculture: urban bias and rural planning in India', in Streeten, P. and Lipton, M. (eds.), *The crisis in Indian planning*, London, 130–47.

LIPTON, M. (1973): 'Urban bias and rural planning in India', in H. Bernstein, *Underdevelopment and development*, London, 235–6.

LIST, F. (1885): *The national system of political economy*, London.

LUYKX, N. (1970): 'The Comilla Project, East Pakistan', in A. H. Bunting (ed.), *Change in agriculture*, International seminar on change in agriculture, University of Reading, London, 361–9.

MABRO, R. (1971): 'Employment and wages in dual agriculture', *Oxf. Econ. Pap.* (NS), 23 (3), 401–17.

MACARTHUR, J. D. (1971): 'Some general characteristics of farming in a tropical environment', in Ruthenberg, H., *Farming systems in the tropics*, Oxford.

McKENZIE, HON. BRUCE (1970): 'Kenya's future agricultural development', *World Crops*, 22 (5), 297–8.

MACKIE, R. B., DAWE, M. T. and LOXLEY, C. F. (1927): *Report of the rice commission on its enquiry into the position of the rice industry*, Sessional paper no. 7 of 1927, Freetown.

MCMASTER, D. N. (1962): *A subsistence crop geography of Uganda*, Bude, Cornwall.

MCMASTER, D. N. (1968): '*Uganda: initiatives in Agriculture*, 1888–1966', *Trans. and Pap. Inst. of Brit. Geogr.*, 44, 241–58.

MCPHERSON, W. W. (1968): 'Status of tropical agriculture', in McPherson, W. W., *Economic development of tropical agriculture*, Gainesville, 1–22.

MAINA, J. W. and MACARTHUR, J. D. (1970): 'Land settlement in Kenya', in A. H. Bunting (ed.), *Change in agriculture*, International seminar on change in agriculture, University of Reading, London, 427–35.

MALONE, C. C. (1970): 'The intensive agricultural districts programme ('package' programme), India', in A. H. Bunting (ed.), *Change in agriculture*, International seminar on change in agriculture, University of Reading, London, 371–80.

MANSHARD, W. (1957): 'Agrarische Organisationsformen für den binnenmarkt bestimmter Kulturen in Waldgürtel Ghanas', *Erdkunde*, 11, 215–24.

MANSHARD, W. (1961): *Die geographischen Grundlagen der Wirtschaft Ghanas*, Wiesbaden.

MANSHARD, W. (1974): *Tropical agriculture*, London.

MANSHARD, W. (1975): 'Geography and environmental science', *Area*, 7 (3), 147–55.

MARCHAND, B. (1973): 'Deformation of a transportation surface', *Ann. Assoc. Amer. Geogr.*, 63 (4), 507–21.

MARSHALL, P. S. (1970): 'Opportunities in Pakistan', *World Crops*, 22 (1), 26–30.

MARX, K. (1889): *Capital: a critical analysis of capitalist production* (the stereotyped edition), translated from the third German edition by S. Moore and E. Aveling and edited by F. Engels, London.

MASEFIELD, G. B. (1972): *A history of the colonial agricultural service*, Oxford.

MAUDE, A. (1970): 'Shifting cultivation and population growth in Tonga', *Journ. Trop. Geogr.*, 31, 57–64.

MAY, J. M. and MCLELLAN, D. L. (1972): *The ecology of malnutrition in Mexico and Central America*, New York.

MINHAS, B. S. (1970): 'Rural poverty, land redistribution and development strategy: facts and policy', *Indian Econ. Rev.*, 5(1), 97–128.

MIRACLE, M. P. (1962): 'African markets and trade in the Copperbelt', in Bohannan, P. and Dalton, G. (eds.), *Markets in Africa* Evanston, Illinois, 698–738.

MIRACLE, M. (1964): *Traditional agricultural methods in the Congo basin*, Food Research Inst., Stanford Univ., California.

MITTENDORF, H. J. and WILSON, S. G. (1961): *Livestock and meat marketing in Africa*, Rome.

MORGAN, W. B. (1953): 'The lower Shire valley of Nyasaland: a changing system of African agriculture', *Geogr. Journ.*, 119 (4), 459–69.

MORGAN, W. B. (1955): 'Farming practice, settlement pattern and population density in South-Eastern Nigeria', *Geogr. Journ.*, 121 (3), 320–33.

MORGAN, W. B. (1959): 'Agriculture in Southern Nigeria' (excluding the Cameroons), *Econ. Geogr.*, 35, 138–50.

MORGAN, W. B. (1969 A): 'Peasant agriculture in tropical Africa', in Thomas, M. F. and Whittington, G. W. (eds.), *Environment and land use in Africa*, London, 241–72.

MORGAN, W. B. (1969 B): 'The zoning of land use around rural settlements in tropical Africa', in Thomas, M. F. and Whittington, G. W. (eds.), *Environment and land use in Africa*, London, 301–19.

MORGAN, W. B. (1973): *Agricultural location in the less developed world*, Inaugural lecture, London.

MORGAN, W. B. and MOSS, R. P. (1970): 'Farming, forest and savanna in Western Nigeria', *Erdkunde*, 24, 71–80.

MORGAN, W. B. and MUNTON, R. J. C. (1971): *Agricultural geography*, London.

MORGAN, W. B. and PUGH, J. C. (1969): *West Africa*, London.

MORTIMORE, M. J. (1975): 'Peri-urban pressures', in Moss, R. P. and Rathbone, R. J. A. R. (eds.), *The population factor in African studies*, London, 188–97.

MOSS, R. P. and MORGAN, W. B. (1970): 'Soils, plants and farmers in West Africa', in Garlick, J. P. (ed.), *Human ecology in the tropics*, Oxford, 1–31.

MYINT, H. (1964): *The economics of the developing countries*, London.

MYINT, H. (1972): *Southeast Asia's economy: development policies in the 1970s*, Harmondsworth.

MYRDAL, G. (1957): *Economic theory and underdeveloped regions*, London.

MYRDAL, G. (1971): *The challenge of world poverty: a world anti-poverty programme in outline*, Harmondsworth.

NICHOLLS, W. H. (1963): 'An "agricultural surplus" as a factor in economic development', *Journ. Polit. Econ.*, 71 (1), 1–29.

NICHOLLS, W. H. (1969): 'The transformation of agriculture in a semi-industrialized country: the case of Brazil', in Thorbecke, E. (ed.), *The role of agriculture in economic development*, New York, 311–78.

NIGERIAN GOVERNMENT (1952): *Report of the sample census of agriculture, 1950–51*, Lagos.

NORCLIFFE, G. B. (1969): 'The role of scale in locational analysis: the phormium industry in St. Helena', *Journ. Trop. Geogr.*, 29, 49–57.

NULTY, L. (1972): *The green revolution in West Pakistan: implications of technological change*, New York.

NURKSE, R. (1953): *Problems of capital formation in underdeveloped countries*, New York.

O'CONNOR, A. M. (1975): 'Sugar in tropical Africa', *Geogr.*, 60 (1), 24–30.

ODELL, P. R. (1974): 'Geography and economic development with special reference to Latin America', *Geogr.*, 59 (3), 208–22.

OGADA, F. (1970): 'Maize in the agriculture of Kenya', *World Crops*, 22 (5), 302–3.

OJALA, E. M. (1969): 'Agriculture in the world of 1975: general picture of trends', in Papi, U. and Nunn, C. (eds.), *Economic problems of agriculture in industrial societies*, Proceedings of a conference held by the International Economic Association, London, 3–25.

OLIVER, H. (1965): 'Irrigation as a factor boosting food and fibre production', *Proc. Nutr. Soc.*, 24, 13.

ORTIZ, S. R. (1973): *Uncertainties in peasant farming*, London.

OXFORD ECONOMIC ATLAS OF THE WORLD, 4th edn. (1972), London.

OYELEYE, D. A. (1973): 'Agricultural landuse in Oyo Division viewed against the background of von Thünen's theory of land utilization', *Journ. Trop. Geogr.*, 37, 39–52.

PANNIKAR, P. G. K. (1961): 'Rural savings in India', *Econ. Dev. and Cult. Change*, 10, 64–85.

PEET, J. R. (1969): 'The spatial expansion of commercial agriculture in the nineteenth century: a von Thünen interpretation', *Econ. Geogr.*, 45 (4), 283–301.

PERROUX, F. (1955): 'Note sur la notion de pôle de croissance', *Economie appliquée*, 8, 307–20, translated by I. Livingstone as 'Note on the concept of "growth poles" ' in Livingstone, I. (ed.), *Economic policy for development* (1971), Harmondsworth, 278–89.

PHILLIPS, T. A. (1965): 'Nucleus plantations and processing factories: their place in the development of organised smallholder production', *Trop. Sci.*, 7, 99–108.

PICKSTOCK, M. (1974): 'Turning point for world agriculture: report on the 1973 FAO Conference', *World Crops*, 26 (1), 38–40.

PINCUS, J. (1968): *The economy of Paraguay*, New York.

PLUMBE, A. J. (1974): *Marketing co-operatives and spatial cross-subsidisation in Tanzania*, Unpubl. paper presented at Inst. Brit. Geogr. (Devg. Areas Study Group) Conference, London.

POLLITT, B. H. (1971): 'Employment plans, performance and future prospects in Cuba', in Robinson, R. and Johnston, P. (eds.), *Prospects for employment opportunities in the 1970s*, University of Cambridge Overseas Study Committee Conference, London.

PORTERES, R. (1955): 'L'introduction du maïs en Afrique', *Journ. d'Agric. Trop. et de Botan. Appl.*, 2, 221–31.

PRED, A. (1967): 'Behaviour and location: foundations for a geographic and dynamic location theory', Part I, *Lund Stud. in Geogr.*, series B, 27.

PREST, A. R. (1970): 'The role of direct taxation', in Taylor, M. C. (ed.), *Taxation for African development*, London, 238–56.

PRESTON, D. A. (1969): 'The revolutionary landscape of highland Bolivia', *Geogr. Journ.*, 135, 1–16.

PRESTON, D. A. (1970): 'New towns: a major change in the rural settlement pattern of highland Bolivia', *Journ. of Lat. Amer. Stud.*, 2, 1–27.

PRESTON, D. A. (1972): *Internal domination: small towns, the countryside and development*, Working Paper 17, Department of Geography, University of Leeds.

PRESTON, D. A. (1973): *Farmers and towns: rural-urban interaction in highland Bolivia*, Working Paper 32, Department of Geography, University of Leeds.

PROTHERO, R. M. (1957): 'Land use at Soba, Zaria Province, Northern Nigeria', *Econ. Geogr.*, 33, 72–86.

RAM, M. (1975): 'Ten years of dwarf rice in India', *World Crops*, 27 (1), 33–6.

RANA, A. S. (1971): 'Introduction and scope for power tillers in Nigeria', *World Crops*, 23 (5), 256–9.

REDFIELD, R. (1941A): *The folk culture of Yucatan*, Chicago.

REDFIELD, R. (1941B): 'The folk society', *Amer. Journ. of Soc.*, 52 (4).

REIDINGER, R. B. (1974): 'Institutional rationing of canal water in Northern India: conflict between traditional patterns and modern needs', *Econ. Dev. and Cult. Change*, 23 (1), 79–104.

RIBEIRO, J. P. and WHARTON, C. R. Jr. (1969): 'The ACAR program in Minas Gerais, Brazil', in Wharton, C. R. Jr. (ed.), *Subsistence agriculture and economic development*, Chicago, 424–38.

RICHARDS, A. I. (1958): 'A changing pattern of agriculture in East

Africa: the Bemba of Northern Rhodesia', *Geogr. Journ.*, 124 (3), 302–14.

RIDDELL, J. B. (1974): 'Periodic markets in Sierra Leone', *Ann. Assoc. Amer. Geogr.*, 64 (4), 541–48.

ROBINSON, G. and SALIH, K. B. (1971): 'The spread of development around Kuala Lumpur: a methodology for an exploratory test of some assumptions of the growth-pole model', *Reg. Stud.*, 5 (4), 303–14.

ROIDER, W. (1970): 'Nigerian farm settlement schemes', in Bunting, A. H. (ed.), *Change in agriculture*, International seminar on change in agriculture, University of Reading, London, 421–6.

ROSCISZEWSKI, M. (1974): 'Organization and typology of socio-economic space in Third World countries', *Norsk Geogr. Tids.*, 28 (1), 41–52.

ROSING, K. E. (1964): 'Nutrition as an index to relative economic development', *Journ. Minnesota Acad. of Sci.*, 32 (1), 43–6.

ROSTOW, W. (1956): 'The take-off into self-sustained growth', *Econ. Journ.*, 66, 25–48.

ROSTOW, W. (1960): *The stages of economic growth; a non-communist manifesto*, Cambridge.

ROUSSEL, L. (1971): 'Employment problems and policies in the Ivory Coast', *Internat. Labour Rev.*, 104 (6), 505–25.

RUTHENBERG, H. (1971): *Farming systems in the tropics*, Oxford.

RUTTAN, V. W. (1968): 'Strategy for increasing rice production in Southeast Asia', in McPherson, W. W. (ed.), *Economic development of tropical agriculture*, Gainesville, 155–82.

SAMUELSON, P. A. (1970): *Economics* (8th edn.), New York.

SAUTTER, G. (1962): 'A propos de quelques terroirs d'Afrique occidentale, essai comparatif', *Etud. rur.*, 4, 24–86.

SCHATZ, S. P. (1965): 'The capital shortage illusion: government lending in Nigeria', *Oxf. Econ. Pap.*, 17 (2), 309–16.

SCHLEBECKER, J. T. (1960): 'The world metropolis and the history of American agriculture', *Journ. Econ. Hist.*, 20 (2), 187–208.

SCHLIPPE, P. DE (1956): *Shifting cultivation in Africa: the Zande system of agriculture*, London.

SCHUH, G. E. (1970): *The agricultural development of Brazil*, New York.

SCHULTZ, T. W. (1953): *The economic organization of agriculture*, New York.

SCHULTZ, T. W. (1964): *Transforming traditional agriculture*, New Haven.

SEAVOY, R. E. (1973): 'The transition to continuous rice cultivation in Kalimantan', *Ann. Assoc. Amer. Geogr.*, 63 (2), 218–25.

SEERS, D. and JOY, L. (ed.) (1971): *Development in a divided world*, Harmondsworth.

SHABTAI, S. H. (1975): 'Army and economy in tropical Africa', *Econ. Dev. and Cult. Change*, 23 (4), 687–701.

SIMPSON, E. S. (1965): 'Discussion of the geographical typology of agriculture', *Agricultural Geography*, 59–74, IGU Symposium of 1964, Dept. of Geography, University of Liverpool.

SINGH, I. and DAY, R. H. (1975): 'A microeconometric chronicle of the Green Revolution', *Econ. Dev. and Cult. Change*, 23 (4), 661–86.

SMITH, M. G. (1965): *The plural society in the British West Indies*, Berkeley.

STAVENHAGEN, R. (1964): 'Changing functions of the community in underdeveloped countries', *Sociologica ruralis*, 4, 315–31.

STERNBERG, H. O'REILLY (1955): 'Agriculture and industry in Brazil', *Geogr. Journ.*, 121, 488–502.

STERNBERG, H. O'REILLY (1967): 'Progrès techniques et décentralisation industrielle dans le paysage rural semi-aride du Nordeste, les problèmes agraires des Ameriques latines', *Colloques internationaux du centre national de la recherche scientifique: sciences humaines* (Paris, 1965), Paris, 251–75.

STERNSTEIN, L. (1967): 'Aspects of agricultural land tenure in Thailand', *Journ. Trop. Geogr.*, 24, 22–9.

STREETEN, P. (1974): 'World trade in agricultural commodities and the terms of trade with industrial goods', in Islam, N. (ed.), *Agricultural policy in developing countries*, London, Conference of the International Economic Association in 1972, 207–23.

SWYNNERTON, R. J. M. (1954): *A plan to intensify the development of African agriculture in Kenya*, Nairobi.

TAAFFE, E. J., MORRILL, R. L. and GOULD, P. R. (1963): 'Transport expansion in underdeveloped countries: a comparative analysis', *Geogr. Rev.*, 53, 503–29.

TAX, S. (1953): *Penny capitalism*, Washington.

TAY, T. H. and WEE, Y. C. (1973): 'Success of a Malaysian agro-industry', *World Crops*, 25 (2), 84–6.

TAYLOR, D. R. F. (1969): 'Agricultural change in Kikuyuland', in Thomas, M. F. and Whittington, G. W. (eds.), *Environment and land use in Africa*, London, 463–93.

TAYLOR, J. A. (1974): 'Current problems in Brazilian coffee production', *K.N.A.G. Geografisch Tijdschrift*, 8 (1), 40–6.

TAYLOR, M. C. (1970): 'The relationship between income tax administrations and income tax policy in Nigeria', in Taylor, M. C. (ed.), *Taxation for African development*, London, 515–33.

THORNTON, D. A. (1972): 'Agricultural development in the Sudan Gezira Scheme', *Trop. Agric.*, 49 (2), 105–14.

TODARO, M. P. (1971): 'Income expectations, rural-urban migration and employment in Africa', *Inter. Lab. Rev.*, 104, 387–413.

TOTHILL, J. D. (ed.) (1948): *Agriculture in the Sudan*, London.

UHLIG, H. (1969): 'Hill tribes and rice farmers in the Himalayas and South-East Asia', *Trans. Inst. Brit. Geogr.*, 47, 1–23.

UNITED NATIONS (1975): *Statistical Yearbook: 1974*, New York.

UNITED NATIONS Food and Agricultural Organization (1963): *Third world food survey*, Rome.

UNITED NATIONS Food and Agricultural Organization (1966): *Production yearbook 1965*, Rome.

UNITED NATIONS Food and Agricultural Organization (1970): *Provisional indicative world plan for agricultural development*, Rome.

UNITED NATIONS Food and Agricultural Organization (1972): *Agricultural adjustment in developed countries*, Rome.

UNITED NATIONS Food and Agricultural Organization (1974): *The state of food and agriculture 1973*, Rome.

UNITED NATIONS Food and Agriculture Organization (1974): *Trade Yearbook 1973*, Rome.

UNITED NATIONS Food and Agricultural Organization (1975): *Monthly Bulletin*, November, 24 (11), Rome.

UNITED NATIONS Food and Agriculture Organization (1975): *The state of food and agriculture 1974*, Rome.

UNITED NATIONS Food and Agriculture Organization (1975): *Production yearbook 1974*, vol. 28.1, Rome.

UPTON, M. (1967): *Agriculture in south-western Nigeria*, University of Reading, Dept. of Agric. Econ., Development Studies No. 3.

URQUHART, A. W. (1963): *Patterns of settlement and subsistence in Southwestern Angola*, Washington, National Academy of Sciences–National Research Council, Office of Naval Research, Report No. 18.

UZUREAU, C. (1974): 'Animal draught in West Africa', *World Crops*, 26 (3), 112–14.

VON NEUMAN J., and MORGENSTERN, O. (1944): *Theory of games and economic behaviour*, Princeton.

VOON PHIN KEONG (1967): 'The rubber smallholding industry in Selangor, 1895–1920', *Journ. Trop. Geogr.*, 24, 43–9.

WAIBEL, L. (1937): *Die Rohstoffgebiete des Tropischen Afrika*, Leipzig (quoted in Manshard (1974), 6–8).

WALDOCK, E. A. and CAPSTICK, E. S. and BROWNING, A. J. (1951): *Soil conservation and land use in Sierra Leone*, Sessional paper no. 1 of 1951, Freetown.

WATTERS, R. F. (1960): 'The nature of shifting cultivation', *Pacif. Viewp.*, 1 (1), 59–99.

WATTERS, R. F. (1966): *Shifting cultivation in Venezuela*, FAO, Rome.

WATTERS, R. F. (1967): 'Economic backwardness in the Venezuelan Andes: a study of the traditional sector of the dual economy', *Pacific Viewp.*, 8 (1), 17–67.

WATTS, D. (1973): 'From sugar plantation to open-range grazing: changes in the land use of Nevis, West Indies, 1950–70', *Geogr.*, 58 (1), 65–8.

WATTS, E. R. (1969): 'Commerce and extension', *World Crops*, 21 (5), 343–5.

WATTS, E. R. (1970): 'A reply to D. H. Humphrey's comment on E. R. Watts, Commerce and extension (1969)', *World Crops*, 22 (2), 115.

WATTS, R. (1973): 'Diversification away from coffee', *World Crops*, 25 (5), 229–30.

WEBSTER, J. B. (1963): 'The Bible and the plough', *Journ. Hist. Soc. Nigeria*, 2 (4), 418–34.

WEE, Y. C. (1970): 'The development of pineapple cultivation in West Malaysia', *Journ. Trop. Geogr.*, 30, 68–75.

WEINTRAUB, S. (1966): *Trade preferences for less-developed countries*, New York.

WHARTON JR., C. R. (ed.) (1969): *Subsistence agriculture and economic development*, Chicago.

WHARTON JR., C. R. (1970): 'The ACAR programme in Minas Gerais, Brazil', in Bunting, A. H. (ed.), *Change in Agriculture*, International seminar on change in agriculture, University of Reading, London, 525–32 (based on a chapter by C. R. Wharton Jr. and J. P. Ribeiro in Wharton Jr., C. R. (ed.) *Subsistence agriculture and economic development*, Chicago, 1969).

WHITTLESEY, D. (1954): 'The regional concept and the regional method', in James, P. E. and Jones, C. F. (eds.), *American geography, inventory and prospect*, Syracuse, 19–68.

WILBANKS, T. J. (1972): 'Accessibility and technological change in Northern India', *Ann. Assoc. Amer. Geogr.*, 62 (3), 427–36.

WILLIAMSON, J. G. (1965): 'Regional inequality and the process of national development: a description of the patterns', *Econ. Dev. and Cult. Change*, 13, 3–45.

WOLF, E. R. (1956): 'Aspects of group relations in a complex society: Mexico', *Amer. Anthrop.*, 58 (6), 1065–78 and in Shanin, T. (ed.), *Peasants and peasant societies* (Harmondsworth, 1971), 50–68.

WOLF JR., C. and WEINSCHROTT, D. (1973): 'International transactions and regionalism: distinguishing 'insiders' from 'outsiders', *Amer. Econ. Rev.*, 63 (2), Papers and proceedings issue, 52–60 and 67–70.

WOLPERT, J. (1964): 'The decision process in a spatial context', *Ann. Ass. Am. Geogr.*, 54, 537–58.

WORLD BANK (1970): *Trends in developing countries*, Washington.

WORLD BANK (1974): *Land Reform*, World Bank paper—rural development series, Washington.

YEATES, M. (1973): 'A pattern of agricultural change in Ethiopia', *World Crops*, 25 (3), 141–44.

Index